Recent Results in Cancer Research 133

P. Pfitzer E. Grundmann (Eds.)

Current Status of Diagnostic Cytology

With 30 Figures and 37 Tables

Springer-Verlag
Berlin Heidelberg New York
London Paris Tokyo
Hong Kong Barcelona
Budapest

Prof. Dr. med., Dr. rer. nat. Peter Pfitzer
Institute for Cytopathology
Center for Pathology
Heinrich Heine University
Moorenstr. 5
40225 Düsseldorf
Germany

Prof. Dr. med. E. Grundmann
Gerhard Domagk Institute for Pathology
University of Münster
Domagkstr. 17
48149 Münster
Germany

ISBN-13: 978-3-642-84953-4 e-ISBN-13: 978-3-642-84951-0
DOI: 10.1007/978-3-642-84951-0

Library of Congress Cataloging-in-Publication Data. Current status of diagnostic cytology / [edited by] P. Pfitzer, E. Grundmann. p. cm. – (Recent results in cancer research : 133) Includes bibliographical references and index. ISBN 3-540-56618-X (alk. paper) : £148.00. – ISBN 0-387-56618-X (alk. paper) 1. Cytodiagnosis. I. Pfitzer, P. (Peter), 1929– . II. Grundmann, E. (Ekkehard) III. Series. [DNLM: 1. Cytology–trends. W1 RE106P v.133 1993 / QY 95 C976 1993] RC261.R35 vol. 133 [RB43] 616.99'4 s – dc20 [616.99'26507582] DNLM/DLC 93-11295

Softcover reprint of the hardcover 1st edition 1993

Typesetting: Best-set Typesetter Ltd., Hong Kong
19/3130/SPS-5 4 3 2 1 0 – Printed on acid-free paper

Preface

Progress in science is often promoted by a new method. Diagnostic cytology, however, developed slowly over a whole century, mainly in differentiating malignant from benign cells from various tissues. The decisive step in this development was the intensive investigation of one localization by Papanicolaou: The application of cytology for screening in the field of gynecologic oncology made it an acknowledged technique. Consequently, materials investigated before were tested again on a larger scale and new ones were included into the program.

The possibility of a wide-range application of this diagnostic method, which carries a low risk for the patient and is low in cost, attracted the specialists of many fields. One of the problems which has resulted is the coordination of training and quality assurance for a large group of people from different fields and with different interests and whose experience in morphology varies.

In this volume general problems of cytology are discussed, as is the question "who is a medical cytologist?" Education and training are the topics of the contributions by Coleman, Holzner, Jenny, Koss and Müller, covering the situation in the European Community, Germany, Austria, Switzerland, and the USA. A special contribution by Lange concerns the situation of cytotechnologists, paramedicals important for cytologic screening programs. Another expansion of diagnostic cytology resulted from new techniques such as DNA cytometry, which has been discussed by Tribukait, the diagnosis of viral infections by Coleman and Wagner, the application of immunocytochemistry by Dalquen et al., AgNOR evaluation by

Rüschoff et al., and the state of the art of automation by Husain. Fine needle aspiration is constantly increasing in application. This technique has made tumors accessible which are not exfoliating cells from an external or internal surface. However, this topic was not included in this volume because a monograph by Koss et al. has been published recently (*Aspiration Biopsy*, Williams & Wilkins, London 1992).

Diagnostic methods are a part of cancer research, and so the *Status of Diagnostic Cytology* has been included in this series. A review of old and new information on applications of diagnostic cytology, training, and the organization are important aspects for an assured quality.

Düsseldorf, Germany
Münster, Germany
July 1993

P. Pfitzer
E. Grundmann

Contents

*List of Contributors**

* The addresses of the authors are given on the first page of each contribution.

[1] Page on which contribution begins.

Current Status of Diagnostic Cytology

L.G. Koss

Montefiore Medical Center, Albert Einstein College of Medicine, Bronx, NY 10467, USA

The field of diagnostic cytology can be divided into three areas: cancer detection, cancer diagnosis, and cancer research. I shall discuss all three of them in sequence.

Cancer Detection

By cancer detection I mean the discovery of occult carcinomas in an asymptomatic population. Cancer detection by cytologic techniques is applicable to a large number of target organs, some very well known and some perhaps somewhat less well known. The primary target, of course, is the uterine cervix, followed by the endometrium, the respiratory tract (lung and larynx) and oral cavity, esophagus, stomach, and the lower urinary tract (flat carcinoma in situ). I shall not discuss the last topic.

Uterine Cervix

The results of a cervix cancer detection program in the Province of British Columbia (Canada) are shown in Table 1. It may be noted that, as a consequence of widespread use of the cervical smear, the rate of invasive cancer of the cervix dropped from 28.4 per 100000 in 1955 to 4.0 in 1987. Similar data were repeatedly reported from other countries, where very well-conducted cytologic screening programs were introduced some years ago, notably Iceland and Finland (Geirsson 1986, Hakama 1978). These surveys are the result of cervix cancer detection programs which have been very well-organized with the help of competent epidemiologists and represent examples of the efficacy of cervical cancer detection within the framework of national health insurance programs. What is striking and important is the fact that in none of the large populations studied until today has there been

Recent Results in Cancer Research, Vol. 133

Table 1. Clinical invasive squamous carcinoma of the cervix: incidence in british columbia. (Courtesy of Dr. George H. Anderson, Vancouver, BC, Canada)

Year	Population over age 20 (In thousands)	Total cases (*n*)	Incidence per 100 000
1955	422.9	120	28.4
1960	486.4	96	19.7
1965	543.2	80	14.7
1970	664.4	82	12.3
1975	805.5	70	8.7
1980	912.9	63	6.9
1985	1063.1	68	6.4
1987	1085.7	44	4.0

Table 2. Results of cytologic screening for cervical cancer of a stable population of approximately 20 000 women 23 years of age and over. (From Marshall 1965)

Year	Positive results confirmed by biopsy	In situ cases	Invasive cases
1958	24	14	10
1959	26	20	6
1960	24	21	3
1961	22	21	1

a complete disappearance of invasive carcinoma of the cervix. There are many reasons for it that will be discussed together with the issue of quality control. In a closely surveyed, repeatedly screened population it is possible to eliminate invasive cancer of the uterine cervix entirely. That was done several years ago in a closed community in the Seattle (Washington, USA) area by Marshall (1965; Table 2). The single case of invasive carcinoma of the cervix observed in this closed population of approximately 20 000 women occurred in a person who was not previously screened and entered the program with delay (Marshall 1965). Thus the goal of elimination of carcinoma of the uterine cervix as a cause of morbidity and mortality is achievable. It remains to be determined whether such a program is cost effective when applied to a large population of women. The fact that inadequate screening may cause the death of women from cervix cancer has been the subject of much soul searching in the United States (Koss 1989). I shall return to this issue in reference to quality control.

With reference to cancer of the uterine cervix, one of the most important scientific questions that needs to be addressed is its precursor states, globally known as cervical intraepithelial neoplasia (CIN). It has been know from many past surveys that the behavior of the precancerous lesions of the uterine cervix is quite unpredictable (Koss et al. 1963; Nasiell et al. 1983, 1986). Some of the lesions will disappear either spontaneously or after biopsies, some of them will persist, often for many years without any change, and some of them will progress either to another form of intraepithelial neoplasia or to an invasive carcinoma. The precancerous states of the uterine cervix behave in a fashion similar to transformed cells in culture. If a cultured cell population is transformed by a virus, one of three events will occur: either the cells will die, or they will redifferentiate into normal cells, or they will become transformed or cancerous. In many ways the behavior of precancerous lesions of the uterine cervix resembles the behavior of transformed cells in culture. While a great deal of speculation is possible, there are no scientific explanations why the behavior of these lesions, and probably precancerous states of other organs as well, is so unpredictable.

Endometrium

The prevalence rate of occult endometrial carcinoma in a cohort of 2586 asymptomatic perimenopausal or postmenopausal women was 6.9 per 1000 (Koss et al. 1984). The study was conducted by direct endometrial sampling techniques, supplemented by routine vaginal and cervical smears. It is of interest that seven of the 12 tumors treated by hysterectomy invaded the myometrium, three superficially and four deeply. The incidence rate was 1.7 per 1000 women years. A concomitant epidemiologic study failed to reveal any statistically valid risk factors, except that early menopause protected women from the disease.

Lung

Another application of a cytologic cancer detection system to lung cancer was based on incidental observations, first published many years ago (review in Koss 1992). Cancer cells of squamous type were observed in the sputum of a handful of patients with negative chest roentgenograms. In some of these patients the lesion was localized by bronchoscopy to one of the secondary bronchi. Either carcinoma in situ or carcinoma in situ with early invasion was observed, suggesting that occult bronchogenic carcinoma may be detected by cytology of the sputum in appropriately selected patients. These observations served as a basis for a large national lung cancer detection study in the United States (Fontana 1986). The national study encompassed approximately 30 000 asymptomatic male cigarette smokers aged 45 or older. The study was conducted in three institutions, the Johns Hopkins University, Baltimore, Maryland; the Mayo Clinic Rochester, Minnesota;

Table 3. Results of the National Cancer Institute (USA) lung cancer detection study[1]

How discovered	Prevalence cases[2] (first screening)		Incidence cases[2] (second and subsequent screenings)	
	(*n*)	(%[3])	(*n*)	(%[3])
Cytology only	37	16.5	58	5.5
Cytology and X-ray	30	13.4	18	1.7
X-ray only or symptoms	156	70.0	983	92.8
	223	100	1059	100

[1] Data compiled with the assistance of Drs. Robert S. Fontana (Mayo Clinic), Myron R. Melamed (Memorial Sloan Kettering Cancer Center), and Melvyn S. Tockman (Johns Hopkins Institutions).
[2] Number of patients with lung cancer.
[3] Percentages approximate.
Participants: 30000 male cigarette smokers, age >50, follow up 5 years.

and the Memorial Sloan Kettering, New York. Unfortunately, the study did not give the expected results (Table 3). The number of lung cancers detected by cytology in the first prevalent screening was very low, inasmuch as cytology detected only 37 (16.5%) of the 223 cases of observed lung cancer. On subsequent screenings (incident cases) only 58 (5.5%) cytologically detected lung cancer cases were discovered out of 1059 lung cancers, most of which were identified either by X-ray or because the patients became symptomatic. The study documented that the detection of lung cancer by sputum cytology was limited to bronchogenic carcinomas of epidermoid type, located mainly in secondary bronchi (Woolner 1981; Woolner et al. 1984). In view of these results, the application of cytologic screening of sputum as a method of lung cancer detection in a large population cannot be recommended. Perhaps in selected very high-risk populations the detection efforts are worthwhile because some of the lung cancers, even those discovered by X-ray, were in an early and resectable state, resulting in a good cure rate (Melamed et al. 1987). Unfortunately, quite a few of these newly detected small carcinomas were of the so-called oat cell type and were not curable.

Oral Cavity

Another aspect of cancer detection by cytologic techniques pertains to cancer of the oral cavity. Cancer of the oral cavity is a fairly common disease, especially in heavy drinkers and smokers. In a survey conducted several years ago (Sandler 1962), it became quite evident that many dentists who usually have the first opportunity to find these carcinomas are not aware

Table 4. Cytologic diagnosis of early oral cancer. (From Sandler 1962)

Patients investigated (*n*)	2758
Total carcinomas (*n*)	315
(%)	100
Clinically unsuspected carcinomas (*n*)	62
(%)	20
Carcinomas in situ (*n*)	28
(%)	9

of the fact that the clinical appearance of many of the superficial, early carcinomas is not suggestive of cancer. Most carcinomas in situ present as areas of redness and not as white lesions or leukoplakia, usually considered to be a suspect lesion. In scrape smears from the red lesions highly abnormal squamous cells may be found, leading to the discovery of a carcinoma in situ in the buccal mucosa. The results of the study conducted several years ago by a dentist (Sandler 1962) are shown in Table 4. In the 2758 patients included in the survey there were 315 carcinomas; 62 of these lesions (20%) were clinically completely unsuspected; among the latter there were 28 carcinomas in situ. Buccal cancer is quite easy to detect, provided that the dentists become aware that most of the important precancerous lesions in the buccal mucosa are not white, as commonly thought, but red in color. In situ carcinoma of the oral cavity is clearly a lesion with a better prognosis than invasive cancer and may be effectively treated.

Esophagus and Stomach

Carcinoma of the esophagus is another example of cancer detection. Cancer of the esophagus is very prevalent in certain areas of China, in the northern littoral of the Caspian Sea, and in Brittany in north-western France (summary in Koss 1992). The Chinese concerned with this natural disaster have initiated a system of cancer detection with esophageal balloons. What is of interest is that in the areas of high prevalence of human esophageal cancer in China, cancer of the gullet in chickens is also observed, for reasons that remain totally obscure (Shu 1985). The esophageal balloon developed by the Chinese was based on a gastric balloon with a rough surface devised by Panico (1952) at New York Hospital, Cornell Medical School, some 30 years earlier. The esophageal balloons can be easily swallowed, as shown by several colleagues of mine who used it in our attempt to duplicate the Chinese results (Greenebaum et al. 1984). A smear is prepared from the outer surface of the balloon. The smears are very similar to cervical smears and appear to be effective in the diagnosis of esophageal carcinoma in situ. Several such smears were shown to me by Dr. Y.-J. Shu who spent about 2 years in our laboratories. Dr. Shu assured that most of these cytologically detected lesions could not be seen by endoscopy. Thus surgical resections of the esophagus were based to a very large extent on cytologic detection and localization of the precancerous lesions. Dr. Shu and some other Chinese scientists claim excellent results of esophageal cancer detection in the high-risk areas of China, resulting in a significant drop in the occurrence of invasive cancer and a major salvage of lives (Shu 1983; Li et al. 1989). It is extremely difficult to get the actual facts of the story: some years ago I travelled to China in an attempt to personally see some examples of precancerous lesions of the esophagus, but wherever I went the lesions were not to be seen. I am not sure what the true status of this cancer detection effort is.

In reference to gastric cancer, a huge detection effort is being conducted in Japan (Fukuda et al. 1967; Yamazaki et al. 1989). Initially, cytologic techniques were extensively used but have now been largely replaced by X-ray and endoscopic techniques. In the historical context it is worth mentioning that the first person to undertake a large-scale gastric cancer detection by means of cytologic techniques was Dr. Rolf Schade when he was active at Newcastle-upon-Tyne in England. He wrote several classical contributions and a book on this topic (Schade 1960).

Cancer Diagnosis

Techniques Based on Exfoliated Cells

In the second part of this overview I will address the issue of cancer diagnosis as the second application of cytologic techniques. I am now addressing the issue of cytologic diagnosis of a clinical abnormality, which is quite different from cancer detection targeting a completely asymptomatic population. The topic must be divided into diagnosis based on exfoliative cytology and diagnosis based on aspiration biopsies.

In reference to cancer diagnosis by exfoliated cells the targets are the female genital tract, lung, effusions, cerebrospinal fluid, the lower urinary tract (high-grade tumors), upper gastro-intestinal tract (oesophagus and stomach), and many other fluids, for example bile. I shall not illustrate any of these because I think the subject is too familiar for me to repeat known facts that have been recently summarized (Koss 1992).

Techniques Based on Aspiration Biopsy

The second cytologic diagnostic technique is the aspiration biopsy by means of a thin needle, sometimes referred to as fine-needle aspiration biopsy (FNAB). The technique can be applied either to palpable lesions such as lymph nodes, salivary glands, thyroid, breast, skin, soft tissues, prostate, testis, or ovary, or to organs that require roentgenological guidance. These include the lung; the mediastinum; the breast (in reference to mammographically detected clinically occult abnormalities); abdominal organs such as the liver, spleen, pancreas, kidney, retroperitoneal masses (lymph nodes and soft tissue); and more recently the central nervous system (Koss et al. 1992).

The difference between cytologic diagnosis and cytologic detection of cancer and precancerous states is quite apparent: instead of a presumptive diagnosis that must be confirmed by biopsy, the aspiration cytologist is a first-line diagnostician on whose judgement major decisions will be made in reference to patients' treatment and prognosis. From early timid attempts at this technique made many years ago the system has evolved into a major diagnostic discipline. In my judgement, anyone who is calling him/herself a cytologist or a pathologist with interest in cytology has to be competent in this area of knowledge because this has become the next diagnostic frontier.

The introduction of contemporary aspiration cytology occurred in 1904 when two British military surgeons, Graig and Grey, working in Uganda used a syringe to aspirate lymph nodes for the diagnosis of sleeping sickness. In an article published simultaneously in Lancet and the British Medical Journal they described their observations as follows: "At first the glands were excised; that was soon found to be unnecessary as it is easy to puncture a superficial gland with a hypodermic needle to suck up some of the juice into the needle and to blow this out on a slide. The actively moving trypanosomes were readily found." In 1930, a paper from Memorial Hospital for Cancer and Allied Diseases known today as the Memorial Sloan Kettering Institute, published by Martin and Ellis addressed the issue of cancer diagnosis by aspiration. Martin decided on the aspiration techniques in the 1920s because the chief of pathology at the Memorial Hospital for Cancer, Dr. James Ewing (after whom the tumor of the bone has been named as the Ewing's sarcoma) objected to an open biopsy of cancer because he thought of the biopsy as contributing to the spread of disease. Martin, being unwilling to treat patients without a confirmatory diagnosis, developed a system of syringe-and-needle aspiration which was crude but very effective in confirming clinical suspicion of cancer. Martin was assisted by Edward Ellis who was Ewing's technician, assistant, and factotum. I heard from many sources that Ellis became an excellent diagnostician without any background either in medicine or in pathology. The fundamental contributions to the contemporary applications of the thin-needle aspiration technique were due to the efforts Lopes-Cardozo (1954) in Holland and Söderström (1966) in Sweden. Franzén et al. (1960), in Sweden, developed the single-grip syringe for the aspiration of the prostate. The technique today encompasses every organ in the body.

Some of the key issues having to do with the application of the aspiration technique have to do with techniques. There are two schools: one is based on air-dried smears and hematologic techniques using May-Grünwald-Giemsa or related stains. The other, based on the tradition of pathology, requires fixation of smears, to be stained either with Papanicolaou stain or with hematoxylin-eosin. There are some benefits to either method. The air-dried smear technique is helpful in visualizing some cytoplasmic features such as secretory granules and some acellular material such as colloid. On the other hand, in a fixed preparation there is a much better preservation of the nuclear structure (Koss et al. 1992).

Other critical issues have to do with the interpretation of the aspiration biopsy smears. The pathologists must learn to translate two-dimensional structures of histology into three-dimensional and dispersed structures represented in smears. To perform such tasks well, a fundamental knowledge of anatomy and pathology of the organs to be examined is essential. Each organ is not only composed of unique cells but also shows an organ-specific arrangement of these cells. The tremendous diversity of normal cytology makes the application of this technique to abnormal samples a true diagnostic challenge.

There are some of important features in the aspiration biopsy that perhaps are not sufficiently stressed or estimated. The presence of cell products such as melanin, psammoma bodies, mucin, or colloid may be extremely helpful in the diagnosis (summary in Koss et al. 1992). For example, in a breast aspirate, the presence of mucus outside otherwise benign-looking cell clusters is diagnostic of a colloid carcinoma of the breast. On the contrary, the presence of mucus within the cytoplasm of small signet ring cells is indicative of lobular carcinoma, as first described by Spriggs and Jerrome (1975). This information can be applied to aspirates of many different organs: for example, the presence of such small cells in an aspirate from an orbital mass in a woman can lead to a secure diagnosis of a metastatic mammary carcinoma. The presence of other cell products may lead to other conclusions. The presence of tyrosine crystals in an aspirate from a tumor of the parotid gland is absolutely diagnostic of a benign mixed tumor or a pleomorphic adenoma (Koss et al. 1992). These examples show that a mastery of a broad variety of targets and their pathology is as essential in diagnostic cytology as it is in surgical pathology.

A very important application of aspiration cytology is in the diagnosis of occult carcinoma of the breast. These lesions can now be aspirated under mammographic guidance using stereotactic techniques. Such aspirates require considerable manual and interpretative skills. The creation of diagnostic breast cancer centers staffed by competent radiologists, competent mammographists, and competent pathologists is probably the way of the future. The technique of aspiration of small breast lesions is not for amateurs. Its correct application may make a great deal of difference in the life of a woman (Azavedo et al. 1989). The evidence suggests that the very small carcinomas of the breast, i.e., less than 1 cm in diameter, may have a favorable outcome.

The technique is also applicable to many infectious disorders. Many of these infections have become extraordinarily important today, mainly because of an increased number of patients with acquired immunodeficiency syndrome (AIDS). These people may be compared to living incubators that support the proliferation of bacterial, parasitic, fungal (opportunistic and pathogenic), and viral agents that often cause their death. Tuberculosis of the breast is a good example of an infectious process. In the case illustrated in Fig. 1, a hard palpable mass in the breast of a woman in her 40s was

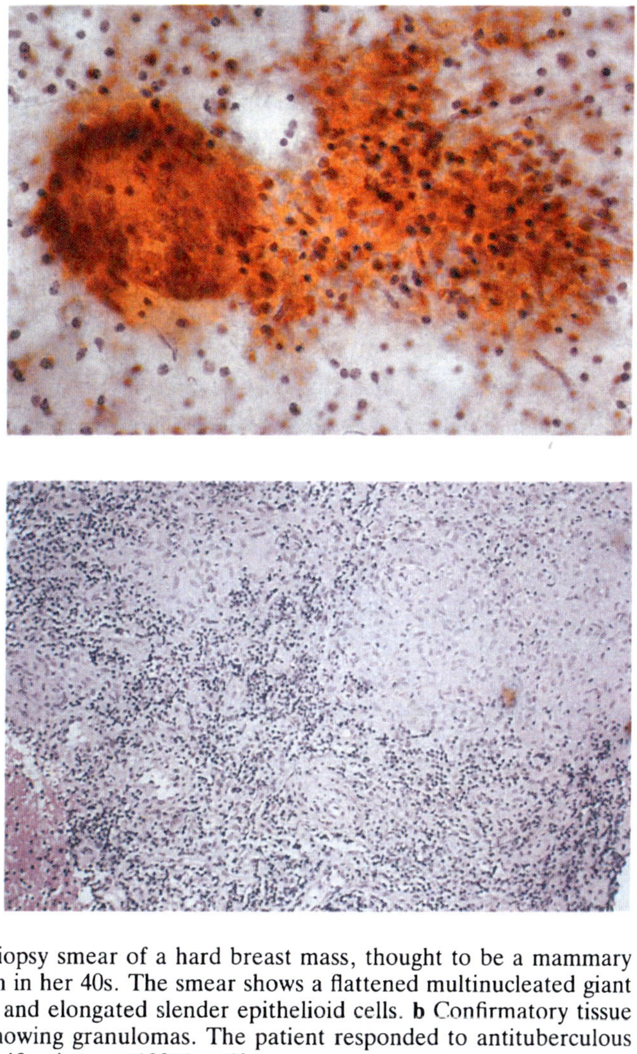

Fig. 1. a Aspiration biopsy smear of a hard breast mass, thought to be a mammary carcinoma, in a women in her 40s. The smear shows a flattened multinucleated giant cell of Langhans' type and elongated slender epithelioid cells. **b** Confirmatory tissue biopsy of the breast showing granulomas. The patient responded to antituberculous therapy. Original magnification **a** ×100; **b** ×40

hought clinically to be a carcinoma. The recognition of the components of a uberculous granuloma may be more difficult in an aspiration smear than in he customary histologic section.

Still, the Langhans' cells can be recognized as multinucleated cells and he epithelioid cells as slender elongated, carrot-shaped cells. A broad 'ariety of other agents may be identified in patients with AIDS ranging from 'iruses such as herpes virus, cytomegalovirus, or polyoma virus to parasites such as toxoplasma. Knowledge of microbiology and parasitology, the norphology of the organisms, and of the lesions induced by them is essential o perform the diagnostic tasks accurately.

The benefits of needle aspiration cytology may be summarized as follows. For palpable lesions it is a simple office procedure that rarely requires inesthesia. It is also a low-cost procedure especially for intrathoracic and ibdominal lesions: an exploratory laparotomy or thoracotomy can be nearly :ompletely avoided as they have been largely replaced by transcutaneous ieedle aspirations. It is a low-risk procedure that the patients generally)refer to a surgical procedure. It is a rapid and reliable procedure in :xperienced hands and it allows for patients' participation and therapeutic lecisions, for example in breast cancer. It also offers research options.

Research Options

The cytologically aspirated sample lends itself to a number of investigative)rocedures. With the availability of a broad spectrum of antibodies, immuno:hemistry has been widely applied and became essential in the diagnosis and :lassification of malignant lymphomas (Tani et al. 1988). Classification of)oorly differentiated tumors may be improved by a panel of monoclonal intibodies to intermediate filaments and cell epitopes (Domagala et al. .987). Space does not permit me to consider molecular biology in this 'olume. Still, it has been shown that gene rearrangement in malignant ymphoma can be documented on the aspirated sample (Lubinski et al. .988). Very recently lymph node aspirates were used to document the)resence of Epstein-Barr virus DNA by polymerase chain reaction in the liagnosis of nasopharyngeal carcinoma (Feinmesser et al. 1992).

Flow cytometry (Table 5) and image analysis (Table 6) represent two)ther important research techniques applicable to the aspirated sample summary in Koss 1982, 1992). Both techniques are suitable for DNA neasurements, quantitation of gene expression, and immunophenotyping of :ell populations using various monoclonal antibodies. Using the technique)f image analysis, the DNA measurements are, in some instances, more iccurate than by flow cytometry (Koss et al. 1989b). Cell cycle analysis, iowever, is much more accurate with flow cytometry. Using the image inalysis techniques, it is possible to quantitate gene expression and measure i number of other parameters (Koss et al. 1989a; Czerniak et al. 1990). For

Table 5. Diagnostic cytology: application of research techniques – flow cytometry

DNA measurements and cell cycle analysis
Quantitation of gene expression
Estimation of proliferative status of cell populations (Ki67, cyclin)
Immunophenotyping

Table 6. Diagnostic cytology: application of research techniques – image analysis

DNA measurements
Measurement of gene expression
Steroid binding
Oncogene expression
Antigenic determinants and epitopes
Morphologic classification
Image analysis
Artificial intelligence

example, every breast cancer should be studied for the presence of steroid receptors, expression of several oncogenes such as HER2/neu and DNA measurements (Bacus et al. 1988, 1989, 1990). No doubt in the future there will be other antigenic determinants that may prove to be of prognostic value. An example of measurements of a protein product of the oncogene *fos* by image analysis and by flow cytometry is shown in Fig. 2. It is important to verify the specificity of the antibody by Western blot technique or other molecular-biologic methods. Clearly this technology is applicable to many other targets and its full value must yet be explored. In recent times we have been studying the performance of a yet different system of cell classification combining image analysis with artificial intelligence. The system has the commercial name Papnet. The system combines an automated microscope with sophisticated image analysis and a neural net, a form of artificial intelligence for the screening of cervical smears. By image analysis the system selects from the hundreds of thousands of cells in a cervical smear about 50000 cell images. From these 50000 cell images the artificial intelligence circuit selects 64 images for display on a screen. The resolution of the system is high. In some cases, the display of abnormal cells side by side has a high teaching value. This system, which, I think, will serve as a quality control apparatus, is currently in industrial production.

Other applications of molecular techniques include in situ hybridization with DNA or RNA (summary in Koss 1992). An example of in situ hybridization of a vulvar condyloma with human papillomavirus, type 11, is shown in Fig. 3. In situ hybridization techniques are applicable to many different targets, including chromosomes in nondividing cells. The clinical value of many of those techniques is likely to grow with time.

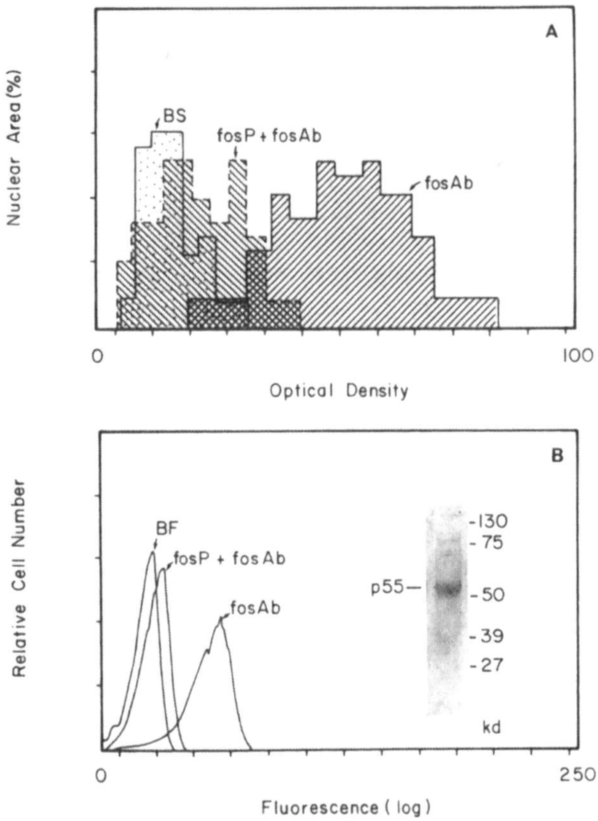

Fig. 2.A,B. Measurements of protein p55, the product of oncogene *fos*, in human breast cancer cell line MCF7-K0. The antibody to p55 was an affinity-purified sheep antibody DCP821 and its complementary 15 amino acid peptide DCX3210 (Cambridge Research Biochemicals). The specificity of the antibody was tested in a Western blot (**B**, *right*). Two methods of measurements are demonstrated by image analysis (**A**) and by flow cytometry (**B**, *left*). The antigen antibody reaction for image analysis was visualized by avidin-biotin complex immunoperoxidase assay and the measurements of the nuclear density were performed on Cell Analysis System CAS 100 apparatus. For flow cytometry the antibody was revealed with fluorescein isothiocynate labelled anti-sheep goat antibody. The measurements were performed on EPICS C flow cytometer (Coulter Electronics). **A** shows the increase of the antibody (fos AB) over control measurements with blocked antibody (*fosP* + *fosAb*) and background staining (*BS*). **B** shows an increase of the antibody expression (*fosAb*) over blocked antibody (*fosP* + *fosAb*) and background fluorescence (*BF*). (Modified from Czerniak et al. 1990)

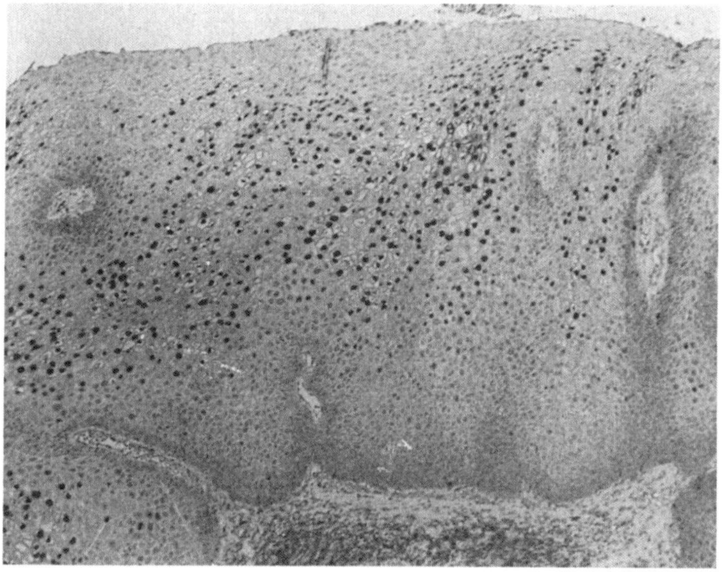

Fig. 3. Vulvar condyloma hybridized with a biotinilated probe to human papillomavirus DNA type II, under stringent conditions. The presence of viral DNA is revealed by black nuclei. ×80

Error Rate and Quality Control

The current problems in diagnostic cytology are shown in Table 7. The error rates must be considered for cancer screening and cancer diagnoses. In reference to screening for cervical cancer a survey was published by van der Graaf et al. (1987) from Dr. Vooijs laboratory. This was a review of 100 based on 165 000 women who received a second smear at an interval of 3 years. There were 555 women whose first smear was negative and whose second smear, 3 years afterwards, revealed a precancerous lesion or cancer. The previous smears from these 555 women were reviewed; 68 were thought to be unsatisfactory (12%), and 60% were thought to show some form of atypia even though the form of atypia was not further specified. In 30% of

Table 7. Current problems in diagnostic cytology

Acceptable error rate
Cancer screening
Cancer diagnosis
Standards with performance
Quality control
Uniform nomenclature (cytopathology – histopathology)

the women (163 out of 555) the previous smear disclosed a neoplastic lesion that was missed on the first screening. That is a very high rate of missed smears. The results may be presented as 163 errors that occurred in a population of 165 000 women reducing it to a miniscule percentage. In my judgement this is not the correct way of presenting the results of screening. It is very hard to pinpoint the screening error rate in other laboratories because such data have not been published. By reviewing prior smears on patients who develop neoplastic lesions, the error rate in screening is probably no less than 5% and possibly more, even in excellent laboratories. The error is significantly greater if smears are used as a follow-up procedure for women with lesions. In at least one third of the cases the second smear will be negative even though on biopsy cervical intraepithelial abnormalities will be present (review in Koss 1992). Reasons for the high false-negative error rate on the repeat second smear remain unknown, and follow up of patients by cervical smears carries with it a very high margin of error. Clearly there is room for improvement in the screening of cervical smears.

Future of European Cytology

In my concluding remarks I should like to address the issue of the future of cytology. What can be done to set the appropriate standards for cytology in Europe? The current problems with diagnostic cytology include training and certification, a professional background, technical problems (background and schooling, and the possibility of limited licensure). I know that many of the people who performed pioneering work in cytology do not necessarily have the kind of background that I personally think is essential for the future. I should like to suggest that, unless appropriately trained pathologists are recruited to the field, there is never going to be first-class cytology in Europe or anywhere else. My own judgement is that professionals qualified to assume responsibilities in this field should have a minimum of 3 years of training and experience in anatomic pathology plus at least 1 year's fellowship in cytopathology under very close supervision. That may appear revolutionary but perhaps the justification of this recommendation will be seen.

The cytotechnologists are an important part of this entire system not only as technicians but also as very important contributors and colleagues without whose help and assistance the system cannot function properly. These valuable people deserve appropriate education, which probably should comprise a university degree or equivalent and at least 1 year of training and 1 year of practice in a dedicated school. They also deserve excellent pay for the efforts because they are extremely valuable members of the team and I know that at this point in time they are not receiving the rewards to which they are entitled. Clearly my remarks represent only a very superficial overview of what has become a very large and important field of medicine and science; however, I hope that they will be helpful in further deliberations.

References

Azavedo E, Svane G, Auer G (1989) Stereotactic fine-needle biopsy in 2594 mammographically detected non-palpable lesions. Lancet 1:1033–1036

Bacus S, Flowers JL, Press MF, Bacus JW, McCarty KS Jr (1988) The evaluation of estrogen receptor in primary breast carcinoma by computer assisted image analysis. Am J Clin Pathol 90:233

Bacus SS, Goldschmidt R, Chin D, Moran G, Wienberg D, Bacus JW (1989) Biologic grading of breast cancer using antibodies to proliferating cells and other markers. Am J Pathol 135:783–792

Bacus SS, Bacus JW, Slamon DJ, Press MF (1990) HER-2/Neu oncogene expression and DNA ploidy analysis in breast cancer. Arch Pathol Lab Med 114:164–169

Czerniak B, Herz F, Wersto RP, Alster P, Puszkin E, Schwarz E, Koss LG (1990) Quantitation of oncogene products by computer-assissted image analysis and flow cytometry. J Histochem Cytochem 38:463–466

Domagala W, Lubinski J, Lasota J, Woyke S, Wozniak L, Szadowska A, Weber K, Osborn M (1987) Decisive role of intermediate filament typing of tumor cells in differential diagnosis of difficult fine needle aspirates. Acta Cytol 30:253–266

Feinmesser R, Miyazaki I, Cheung R, Freeman JL, Noyek AM, Dosch HM (1992) Diagnosis of nasopharyngeal carcinoma by DNA amplification of tissue obtained by fine-needle aspiration. N Engl J Med 326:17–21

Fontana RS (1986) Screening for lung cancer: recent experience in the United States. In: Hansen HH (ed) Lung cancer: basic and clinical aspects. Nijhoff, Boston, pp 91–111

Franzén S, Giertz G, Zajicek J (1960) Cytological diagnosis of prostatic tumours by transrectal aspiration biopsy: a preliminary report. Br J Urol 32:193–196

Fukuda T, Shida S, Takita T, Sawada Y (1967) Cytologic diagnosis of early gastric cancer by the endoscope method with gastrofiberscope. Acta Cytol 11:456–459

Geirsson G (1986) Organization of screening in technically advanced counties: Iceland. In: Hakama M, Miller AB, Day NE (eds) Screening for cancer of the uterine cervix. International Agency for Research on Cancer, Lyons, pp 239–250

Graig EDW, Gray ACH (1904) Note on the lymphatic glands in sleeping sickness. Br Med J 1:1252 and Lancet 1:1570

Greenebaum E, Schreiber K, Shu Y-J, Koss LG (1984) Use of esophageal balloon in the diagnosis of carcinomas of the head, neck, and upper gastrintestinal tract. Acta Cytol 28:9–15

Hakama M (1978) Mass screening for cervical cancer in Finland. In: Miller AB (ed) Screening in cancer: a report of a UICC international workshop, Toronto, April 1978. International Union Against Cancer, pp 93–107

Koss LG (1982) Analytical and quantitative cytology: a historical perspective. Anal Quant Cytol Histol 4:251–256

Koss LG (1989) The Papanicolaou test for cervical cancer detection. A triumph and a tragedy. JAMA, 261: 737–743

Koss LG (1992) Diagnostic cytology and its histopathologic bases, 4th edn. Lippincott, Philadelphia

Koss LG, Stewart FW, Foote FW, Jordan MJ, Bader GM, Day E (1963) Some histological aspects of behavior of epidermoid carcinoma in situ and related lesions of uterine cervix. A long-term prospective study. Cancer 16:1160–1211

Koss LG, Schreiber K, Oberlander SG, Moussouris HF, Lesser M (1984) Detection of endometrial carcinoma and hyperplasia in asympotomatic women. Obstet Gynecol 64:1–11

Koss LG, Czerniak BH, F, Wersto RP (1989a) Flow cytometric measurements of DNA and other cell components in human tumors: a critical appraisal. Hum Pathol 20:528–548

Koss LG, Wersto RP, Simmons DH, Deitch D, Freed SZ (1989b) Predictive value of DNA measurements in bladder washings. Comparison of flow cytometry image cytophotometry and cytology of voided urine in urothelial tumors. Cancer 64: 916–924
Koss LG, Woyke S, Olszewski W (1992) Aspiration biopsy. Cytologic interpretation and histologic bases, 2nd edn. Igaku Shoin, New York
Li J-Y, Ershow AG, Chen Z-J, Wacholder S, Li G-Y, Guo W, Li B, Blot WJ (1989) A case-control study of cancer of the esophagus and gastric cardia in Linxian. Int J Cancer 43:755–761
Lopes-Cardozo P (1954) Clinical cytology. Stafleu, Leiden
Lubinski J, Chosia M, Huebner K (1988) Molecular genetic analysis in the diagnosis of lymphoma in fine needle aspiration biopsies: I. Lymphoma versus benign lymphoproliferative disorders. Anal Quant Cytol Histol 10:391–398
Marshall CE (1965) Effect of cytologic screening on the incidence of invasive carcinoma of the cervix in a semi-closed community. Cancer 18:153–156
Martin HE, Ellis EB (1930) Biopsy by needle puncture and aspiration. Ann Surg 62:169–181
Melamed MR, Flehinger BJ, Zaman MB (1987) Impact of early detection on the clinical course of lung cancer. Surg Clin North Am 67:909–924
Nasiell K, Nasiell M, Vaclavinkova V (1983) Behavior of moderate cervical dysplasia during long-term follow-up. Obstet Gynecol 61:609–614
Nasiell K, Roger B, Nasiell M (1986) Behavior of mild cervical dysplasia during long term follow-up. Obstet Gynecol 67:665–669
Panico FG (1952) Cytologic patterns in benign and malignant gastric and esophageal lesions. Surg Gynecol Obstet 94:733–742
Sandler HC (1962) Cytological screening for early mouth cancer. Interim report of the Veterans Administration Cooperative Study of oral exfoliative cytology. Cancer 15:1119–1124
Schade ROK (1960) Gastric cytology. Principles, methods, and results. Arnold, London
Shu Y-J (1983) Cytopathology of the esophagus. An overview of esophageal cytopathology in China. Acta Cytol 27:7–16
Shu YJ (1985) Cytopathology of esophageal cancer. Masson, New York
Söderström N (1966) Fine-needle aspiration biopsy: used as a direct adjunct in clinical diagnostic work. Almqvist and Wiksell, Stockholm
Spriggs AI, Jerrome DW (1975) Intracellular mucous inclusions. A feature of malignant cells in effusions in the serous cavities, particularly due to carcinoma of the breast. J Clin Pathol 28:929–936
Tani E, Christensson B, Porwit A, Skoog L (1988) Immunocytochemical analysis and cytomorphologic diagnosis on fine needle aspirates of lymphoproliferative disease. Acta Cytol 32:209–215
Van der Graaf Y, Vooijs GP, Gaillard HLJ, Go DMDS (1987) Screening errors in cervical cytologic screening. Acta Cytol 31:434–438
Woolner LB (1981) Recent advances in pulmonary cytology: early detection and localization of occult lung cancer in symptomless males. In: Koss LG, Coleman DV (eds) Advances in clinical cytology, vol 1. Butterworths, London, pp 95–135
Woolner LB, Fontana RS, Cortese DA et al. (1984) Roentgenographically occult lung cancer: Pathologic findings and frequency of multicentricity during a 10 year period. Mayo Clin Proc 59:453–466
Yamazaki H, Oshima A, Murakami R, Endoh S, Ubukata T (1989) A long-term follow-up study of patients with gastric cancer detected by mass screening. Cancer 63:613–617

Current Status of Diagnostic Cytology in Germany

H.-A. Müller

Pathologisches Institut der Universität Würzburg, Joseph-Schneider-Str. 2, 97080 Würzburg, Fed. Rep. of Germany

The present status of diagnostic cytology in Germany is characterized by the imbalance of practical and intellectual capacities and, on the other hand, their circumspect and conscientious application. The discrepancy is caused by the peculiarities of professional and administrative conditions in our country.

The professional problems in Germany include the fact that no clearly defined or approved profile of the cytologist or cytopathologist exists. Nine years ago extensive interprofessional discussion led to the rejection of "cytology" as a partial or additional specialty by gynecologists as well as pathologists. Among other reasons both professional groups seemed to fear undesirable competition by autonomous cytologists. Up to now that situation has remained unchanged, especially after the 94th "Deutscher Ärztetag" (Conference of German Doctors) in 1990.

To put it plainly, in Germany the term of cytologist or cytopathologist is applied to whoever cares to assume it, and there is no professional representation of these professionals. The German Society of Cytology, founded in 1960, is merely a scientific association with an exclusively non-profit-making character. The Society is devoted to the promotion of scientific exchange of experience and opinions in the field of cytology, especially cytodiagnostics; it supports postgraduate and continuing education in cytodiagnostics and in cytology research. The regular members are professionals in medicine, dentistry, veterinary medicine, and the scienes whose activities cover basic research, diagnostics, and marginal fields in cytology.

Among the physicians in this society, gynecology, pathology, internal medicine, and laboratory medicine are the predominant specialties; thus, the German Society of Cytology (DGZ) cannot act as a competent partner in questions of professional policy and is unable to approach the relevant problems.

The core of all the problems in professional policy is seen in the acquisition of cytological skills, continuing education, and, last but not least,

Recent Results in Cancer Research, Vol. 133

the financial outlook of the competent cytologist. Only the regulations of continuing education for the pathologist foresee the acquisition of cytologic skills; the specialist examination requires the attestation of 6000 gynecologic and 4000 nongynecologic slides. The regulations for continuing education in gynecology and obstetrics are supplemented as follows: "Separate attestation of cytologic knowledge and experience is conferred upon evaluation of exfoliative cytology of a minimum of 6000 slides, and/or the additional evaluation of aspiration or punch biopsies in at least 600 slides" (Richtlinien 1988).

In addition to their approved discipline and after passing a practical examination set by the DGZ, other specialists interested in cytology may obtain a certificate attesting their skill in gynecological exfoliative cytology; however, such certificates are validated neither by the federal nor by the regional public health authorities.

No comparable regional or supraregional mode of examination and attestion of knowledge in non-gynecological cytology (e.g., broncho-pulmonary or urogenital cytodiagnostics) has come to my notice so far. It may be emphasized that the chamber of physicians in Saxony is going to formulate and present guidelines for the practice of non-gynecological cytology (R. Goertchen, personal communication).

Obviously, the scope and content of cytodiagnostics in the specialist training of gynecologists and pathologists has been clearly defined by regulations, but serious impairments affected the implementation of these rules in the institutes of professional education, and the problems persist to this day.

Among the pathologists of past decades, many, including the specialists, had a somewhat complicated and even disturbed association with cytology and cytodiagnostics. It may even be said that cytology was misunderstood and even despised by many pathologists. Consequently, many pathologists were educated in academic institutes where cytology played a minor and inadequate role. On the other hand, a good number of prominent pathologists would identify and castigate this misinterpretation, and call for reorientation among their colleagues (Lennert 1973; Holle 1987). In 1973, Karl Lennert, then president of the DGZ, said in his opening address to the 57th meeting of the German Society of Pathology that a specialist in cyto-pathology should belong to every big institute of pathology, and that each university institute of pathology should have its own division of cytopathology with a specialist chair. Despite these directives, no structural integration of cytology into the institutes of pathology took place, apart from some rare exceptions. Most institutes relied solely on the personal concern and diligence of cytologists on their staff who, in fact, provided a kind of figleaf for institutional nudity.

An attempt to arouse a new impetus for cytology, started in 1984 by creating a special working group within the framework of the German Society of Pathology, obviously came too late. After a successful start and first promising developments, the tentative activities failed 3 years later. The initiators could not find any younger successors to take over. In the face of discouragement, leading pathologists would again express their regretful concern, some looking for (simplified) reasons in the scarcity of recognised cytologists, others in more complex problems. Altmann, for instance, took to task not only the incompetence of the superior bodies responsible for the distribution of financial means, but conferred a large part of the guilt on the pathologists themselves.

> Enhancing their virtuous scepticism against new methods, procedures, and modes of investigation, they have hardly ever shown an open, receptive approach, but rather a stubborn refusal. Let me mention the contempt for fine needle biopsy, relegated to clinicians, and the almost complete disregard for Papanicolaou's cytodiagnostics to which the pathologists, caught in traditional experience, were conceding hardly a test function, let alone a decisive property, failing to contribute essentially to its development. (Altmann 1990)

Lack of competence and vision in the administrative authorities has more or less strangulated the establishment, application and progress of general and diagnostic cytopathology in academic institutes of pathology. Cytopathology, as manifested in diagnostics, research, and academic tuition will thus hardly offer an attractive prospect to young physicians interested in pathomorphological problems. Without fundamental changes, cytopathology is hardly likely to last for more than the next generation, provided there are still some competent representatives to be found.

In fact, the precarious situation of cytodiagnostics has not only been caused by the general conditions of professional policy outlined above, but in particular by the peculiarities of the German public health system, especially by the influence of the professional associations of the national health service (KV). This association, an outstanding accomplishment of our public health system, has formulated distinct requirements in the cytology guidelines of 1980 regarding the scope and training of pathologists, gynecologists, and laboratory doctors who want their cytodiagnostic services to be settled by the insurance companies (Richtlinien 1980).

The KV regulations were certainly meritorious, but they failed to prevent uncontrolled or unmerited access of doctors to the practice of cytodiagnostics. Moreover, these guidelines were not intended and were inadequate to secure the quality of such cytodiagnostic services for the future. This recently topical problem ultimately remained unsolved, despite the serious efforts by the professional bodies involved, the victorious opponents apparently being recruited among their own ranks. The KV seems to act like a cartel in the care of its members.

Well aware of the various deficiencies in cytodiagnostic services – such as discount and commission cytology – the KV usually adheres to the strategy of permitting the practice of cytodiagnostics (especially for cancer prophylaxis) only to approved doctors, simultaneously prohibiting such practice to (academic) institutes and clinics involved in professional tuition and training. This system implies a dissipating decentralization of cytodiagnostic activity which leads to a large number of cytologists who have performed only a minimal number of slide examinations.

The system is maintained despite the fact that the prescribed sessions of professional training required by KV become hampered or impossible, so that the professional training of future specialists is impaired as well as that of future clinical and academic teachers. Another problem is the absence of alternative institutions for such training, since the megalaboratories sanctioned by the KV are unable, as a rule, to take their share in postgraduate or later training. The latter failure may be due to practitioners already in the field trying to avoid any potential competition from their own students – quoting the comments of a colleague at the meeting of the DGZ in Freiburg in 1989. With the introduction of our pluralistic health system in the new states of the Federal Republic, a corresponding development has started, initiating the decentralization of cytodiagnostic services at full speed.

As shown above, the professional training ultimately leading to a specialist with cytomorphological orientation is at cross-purposes with the practice of KV-authorized specialists, and no satisfactory solution is in sight. The discrepancies between professional and insurance legislagtion have been formulated once more in this year's doctors' conference in sincerely imploring words: "Professional law regulating practice and specialization must always predominate insurance or KV law" (Ninety-third Deutscher Ärztetag 1990; Ninety-fourth Deutscher Ärztetag 1991).

The climate of health policy in the field of cytodiagnostics is influenced not only by the KV as an institution, but also by the KV acting as a regional factor in administrative decisions. For instance, the acceptance of the new collective agreement by the regional KV, and the implementation of its rules in medical practice has become a paradigm of multiplicity and indefiniteness, and we may well ask whether this is a result of federalism or arbitrariness.

We conclude that the general status of cytodiagnostics in Germany might be considerably better and enjoy a better reputation were it not for the lack of vision, but also for the hypocritical inefficiency of administrative authorities and the tolerated unscrupulous behavior of certain opportunists. The situation is indeed serious, but not completely without hope.

References

Altmann H-W (1990) Die Pathologie an der Schwelle des neuen Jahrhunderts. Würzburger Medizinhistor Mitt 8:351–368

Holle G (1987) Diagnostische Zytologie – Zytopathologie. Pathologe 8:132–136

Lennert K (1973) Begrüßungsansprache. Verh Dtsch Ges Pathol 57:24–26

Ninety-fourth Deutscher Ärztetag (1991) Entschließungen zum Tagesordnungspunkt III. Dtsch Ärztebl 88:A1858–A1860

Ninety-third Deutscher Ärztetag (1990) Weiterbildung soll Pflicht werden. Dtsch Ärztebl 87:B1316–B1318

Richtlinien der Kassenärztlichen Bundesvereinigung über die Voraussetzungen zur Durchführung von zytologischen Untersuchungen im Rahmen der Krebsfrüherkennungsmaßnahmen bei Frauen (Zytologie-Richtlinien) (1980) Dtsch Ärztebl 77:154

Richtlinien über den Inhalt der Weiterbildung in Gebieten, Teilgebieten und Bereichen (1988) Bayer Ärztebl 43:1–52

I. New Methods in Diagnostic Cytology

Tumor Biology in Diagnostic Cytology: DNA Cytometry in Carcinomas of the Bladder and Prostate

B. Tribukait

Department of Medical Radiobiology, Karolinska Institute, Box 60212, S-10401 Stockholm, Sweden

Introduction

The description of nature in quantitative terms is an important step towards a better understanding of the phenomena behind that what the eye is able to recognize.

During the last two decades a rapidly increasing arsenal of methods has become available in histopathology and cytology to measure various cellular components and also to quantify the expression of functional properties of cells and tissues. The aim of this paper is to discuss some aspects of tumor biology based on results of flow cytometric DNA measurements in carcinomas of the bladder and the prostate.

The advantages of flow cytometry are:

1. High speed giving statistical confidence in the DNA histograms measured – usually 30 000 cells are measured in about 1 min. This method is therefore well suited in clinical routine.
2. High resolution, making it possible to define very well the ploidy level of a cell population corresponding to the total amount of chromosomes.
3. By measuring the proportion of cells in S-phase, a functional property of the tumor and an important expression of malignancy can be obtained.
4. Flow cytometry is an objective method and therefore facilitates the exchange of information between various institutions. Flow cytometry, which is a typical cytological method, is, however, non-selective, measuring all cells of a sample and, therefore, only supportive to the morphological evaluation of a tumor.

Recent Results in Cancer Research, Vol. 133

Material and Methods

The investigations were performed on cell material from untreated patients with bladder and prostate carcinomas. Surgical biopsies from bladder carcinomas were mechanically dispersed and fixed in 96% ethanol. Fine-needle aspiration biopsies of the prostate obtained by the transrectal method as described by Franzén et al. (1960) were immediately fixed in 96% ethanol. After pepsin and ribonuclease treatment, the suspensions of single cell nuclei were stained with ethidium bromide. The DNA content of each cell nucleus was analyzed by a flow cytometer. Human lymphocytes were used as an external standard, the coefficient of variation of which was less than 3%. The results were presented as DNA histograms. Unimodal distributed histograms with the same peak position as lymphocytes, or with peaks deviating less than 10% from the position of the peak of the lymphocytes, were defined as diploid. Tumors with gross aneuploid chromosomal aberrations showing one or several additional aneuploid distinct peaks were classified as aneuploid. Aneuploid cell populations with a tetraploid amount of DNA were considered when the cells of the tetraploid peak exceeded 8% and also showed a corresponding G_2-peak at octoploid position. Details of the methods for preparation and staining of the cell material as well as technical aspects of flow cytometry and typical DNA histograms have been described in detail previously (Tribukait 1987). Tumor stages of the bladder and the prostate were assessed according to the recommendations by the UICC (Hermanek and Sobin 1987; Harmer 1978). To assess tumor grade in the bladder the classification adopted by the World Health Organization was used (Mostofi et al. 1973). Based on the degree of nuclear atypia the prostate carcinomas were subdivided into well-, moderately and grossly differentiated according to Esposti (1971).

Results

Bladder Carcinoma

Like many solid tumors, bladder carcinomas can be subdivided into near diploid and grossly aneuploid tumors. Grade 1 tumors are mostly diploid and grade 3 tumors mostly aneuploid. The intermediate large group of grade 2 tumors can be subdivided into about equal groups of diploid and aneuploid tumors. By studying a large number of aneuploid tumors, two distinctly different tumor types can be distinguished: (a) aneuploid grade 3 tumors in which the ploidy values of the majority of the cell populations are distributed in the triploid to tetraploid region with relatively few tumors with tetraploid amount of DNA; and (b) aneuploid grade 2 tumors, the majority of which have tetraploid amounts of DNA (Fig. 1). The first one is log-normally and the second one exponentially distributed in a linear–logarithmic plot

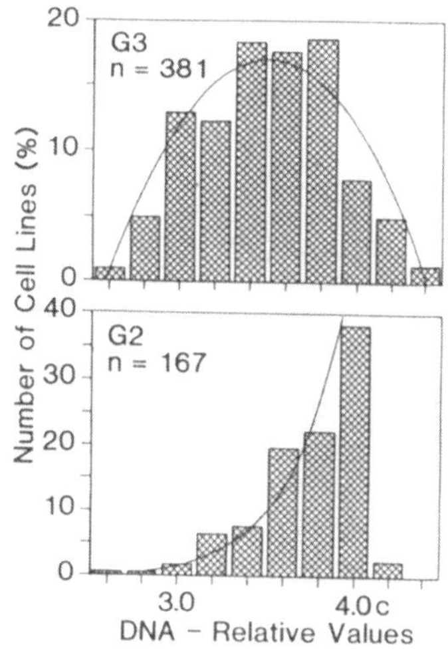

Fig. 1. Distribution of ploidy values of 381 newly detected triploid to tetraploid aneuploid grade 3 tumors and 167 aneuploid grade 2 tumors of the bladder

(Fig. 2). This difference in distribution strongly suggests the existence of two distinct types of aneuploid tumors in the bladder, also with different pathways of development as already discussed by Koss (1979) on the bases of bladder mapping and related histopathologic and clinic observations.

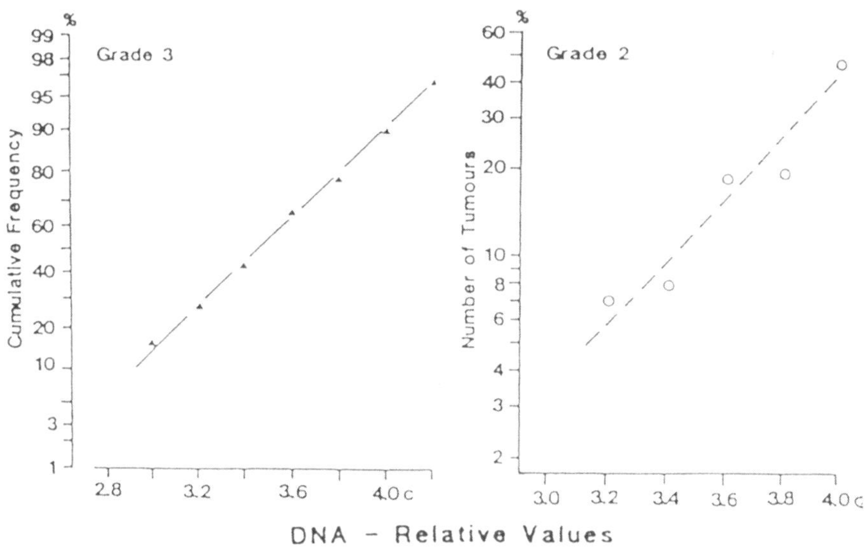

Fig. 2. *Left*, cumulative frequency distribution of ploidy values of triploid to tetraploid aneuploid grade 3 tumors of the bladder plotted on a normal distribution graph paper. *Right*, distribution of ploidy values of triploid to tetraploid aneuploid grade 2 tumors of the bladder on a linear–logarithmic scale

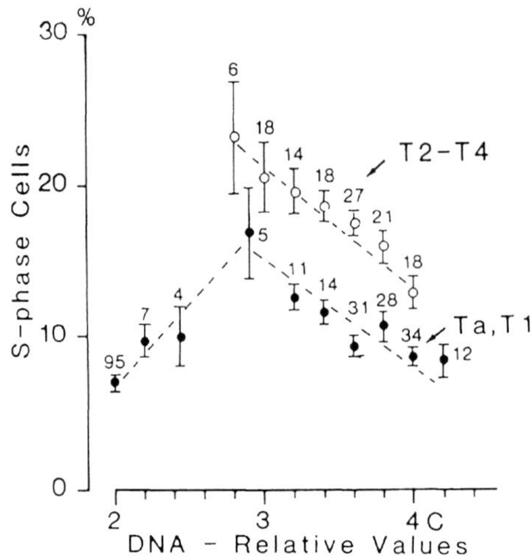

Fig. 3. Percentage of S-phase cells of superficial Ta, T1, and muscle-invasive T2–T4 tumors of the bladder in relation to ploidy of the tumors. Mean values ± SEM and number of tumors

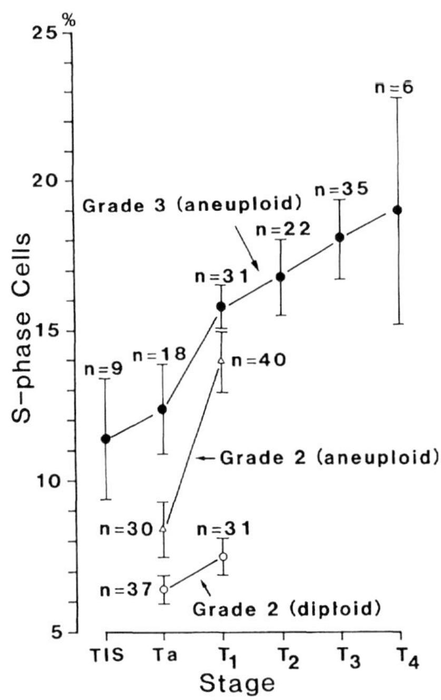

Fig. 4. Percentage of S-phase cells of aneuploid grade 3 (*solid circles*), aneuploid grade 2 (*triangles*), and diploid grade 2 (*open circles*) tumors of the bladder in relation to tumor stage. Mean values ± SEM and number of tumors

Further support for principal differences of these two tumor types are coming from studies of tumor proliferation as measured by the proportions of cells in S-phase and by relating these to ploidy (Fig. 3). Independently of the extent of invasiveness, the proportions of cells in S-phase increase when tumors deviate in their ploidy from tetraploid towards triploid DNA content. However, depending on the existence of invasiveness into the muscle or not, there is a distinct difference between superficial (mostly grade 2) and muscle-invasive (mostly grade 3) tumors.

The significance of proliferation can be further observed from the relationship between the proportions of S-phase cells and stage (Fig. 4). Among aneuploid grade 3 tumors, the proportions of S-phase cells increase continuously from the lowest values in the carcinoma in situ to highest values in the muscle invasive tumors T2–T4. The most pronounced difference in the proportion of S-phase cells is found between noninvasive Ta tumors and T1 tumors invading the subepithelial connective tissue. This difference in the proportions of S-phase cells between Ta and T1 tumors is also found in aneuploid as well as diploid grade 2 tumors.

In conclusion, ploidy in bladder carcinomas is linked to tumor grade, while the proportion of S-phase cells is connected with tumor stage. Aneuploid tumors of grade 2 differ from those of grade 3 by their mode of development and their expression of malignancy as demonstrated by distinct differencies in the rates of proliferation.

Prostate Carcinoma

Adenocarcinoma of the prostate has become one of the most common, and many times *the* most common malignancy in western countries. For example, the age-adjusted incidence rate of 107 per 100 000 men in Sweden in 1988 now exceeds the incidence rate of carcinoma in the female breast and is the highest in this country. This high, and during the last decades increasing, incidence rate of prostate carcinoma is a substantial problem for the health care system and calls for clear strategies in treatment decisions. However, the treatment of prostate cancer is controversial, particularly in the case the localized tumor. The course of the disease is often slow and androgen deprivation usually results in tumor regression, although rarely cure. Therefore, one phalanx of urologists believes that active surveillance, i.e., delayed therapy until tumor progression, is the best treatment alternative. The other phalanx is the advocate of active treatment with radical prostatectomy or radiation therapy.

In this situation studies have focused on the better understanding of the nature of prostate carcinoma, and the development of predictive tests to assess the malignant potential of the prostate carcinoma. In order to contribute to therapeutic decisions, predictive studies have to be performed in cell material from the prostate before therapy has been initiated. The technique of fine-needle aspiration biopsy fulfills this prerequisite and allows the diagnosis and classification of a prostate carcinoma. This technique is well established in Sweden but is still being debated in other countries.

Aspiration of a tumor of the prostate results in high yields of tumor cells. It has been found that the proportion of tumor cells increased from about

Table 1. Frequency of diploid, tetraploid and aneuploid cell populations in fine-needle aspiration biopsies of patients with benign prostate lesions and untreated prostate carcinomas as related to cytologic grade

	Diploid		Tetraploid		Aneuploid		Total
	(*n*)	(%)	(*n*)	(%)	(*n*)	(%)	(*n*)
Benign	486	96	11	2.2	8	1.6	505
Well differentiated	199	58	110	32	36	10	344
Moderately differentiated	113	29	144	37	129	34	386
Poorly differentiated	13	10	31	23	56	67	136

65% in well-differentiated tumors to 85% in poorly differentiated tumors. The proportion of inflammatory cells was about 4% (Wang et al. 1992). Thus, the high yield of tumor cells with a low proportion of benign cells makes cell material obtained by fine-needle aspiration biopsy particularly useful for flow cytometry.

In 505 patients with cytologically benign lesions 96.2% of the samples were diploid. In 2.2% of these patients tetraploid cell populations were observed, and in eight patients (1.6%) clear aneuploid cell populations were found (Table 1).

In tumors, diploid cell populations were found in about 60% of patients with well-differentiated tumors but only in about 10% of patients with poorly differentiated tumors. Aneuploid cell population increased from 10% in well-differentiated to about 70% in poorly differentiated tumors. The large group of moderately differentiated tumors could be subdivided into about equal groups of diploid, tetraploid and aneuploid tumors (Table 1). In summary, morphologically defined tumors can be further characterized in an objective way according to their gross chromosomal composition.

The relationship between ploidy of the tumors and tumor stage is shown in Table 2. Obviously, progression in stage is associated with changes in

Table 2. Frequency of diploid, tetraploid, and aneuploid cell populations in fine-needle aspiration biopsies of patients with untreated prostate carcinomas in relation to clinical tumor stage

Stage	Diploid		Tetraploid		Aneuploid		Total
	(*n*)	(%)	(*n*)	(%)	(*n*)	(%)	(*n*)
1	27	80	5	14	2	6	34
2	51	34	72	49	25	17	148
3	33	12	129	48	105	40	267
4	1	2	18	36	32	62	51

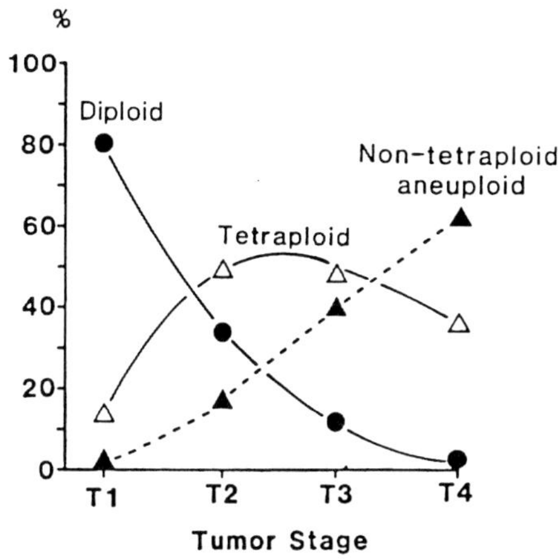

Fig. 5. Frequency of diploid, tetraploid, and non-tetraploid aneuploid cell populations from 500 patients with untreated prostate carcinomas. The curves are best fits calculated according to a three-compartment system

ploidy. Diploid tumors are most common in stage T1 but rarely found in stage T4. The opposite is true for aneuploid tumors, which are rarely found in stage T1 but are most common in stage T4. Tetraploid tumors increase to maximum values in stages T2 to T3.

The typical behavior of these changes in ploidy pattern with tumor stage can be better explained by a graph as shown in Fig. 5. The decrease of diploid tumors and the increase of aneuploid tumors with stage appear to be exponential while tetraploid tumors increase to maximum at stages T2 to T3. This pattern is analogous to that one of a three-compartment system as first described by Bateman (1910) for the decay of radioactive substances. In such a system the decay of the first compartment is a simple exponential function; the second compartment passes through a maximum; and the third compartment, after an induction period, increases exponentially.

For the description of the changes of ploidy with tumor stage we therefore assumed a three-compartment system and calculated according to this mathematical model the transformation rates of diploid and tetraploid tumors. At an assumed interval of 5 years between the different tumor stages, calculated transformation rates for diploid and tetraploid tumors are 17% and 8.5% per year, respectively. As can be seen from Fig. 5, the gradual decrease in the fraction of diploid tumors with stage is nearly exponential as the modal suggests. The rapid initial increase in the fraction of tetraploid tumors and the more gradual decrease in later stages as the tumors are transformed into non-tetraploid aneuploid forms is correctly described. Finally, the accumulation of non-tetraploid aneuploid tumors with stage is also well predicted.

The very good fit of this model to the clinical data is consistent with the transformation from low-grade to high-grade malignancy according to the pathway: near-diploid–tetraploid–non-tetraploid aneuploid. Direct observations in untreated patients during follow-up support such a development of the prostate carcinoma. During a mean observation time of 50 months of 167 patients with low-grade, low-stage tumors, half of the patients with diploid and tetraploid or aneuploid tumors showed local progression after 58 and 40 months, respectively (Adolfson et al. 1990). Repeated fine-needle aspiration biopsies in 84 of these patients after a median time of 47 months revealed changes to higher grade or ploidy in 23% and 24%, respectively (Adolfson and Tribukait 1990).

An important consequence of the assumption of a three-compartment system for prostate carcinoma is the coexistence of several tumor cell populations. Coexistence of diploid with tetraploid and aneuploid tumor cell lines is also the rule in prostate carcinoma according to single-cell DNA measurements of cytologically identified tumor cells. From the view point of therapy, tumor heterogeneity has been discussed as one of the major difficulties and the reason for the failure of chemotherapy. In prostate carcinoma, coexistence of androgen-dependent and -independent cell populations can be expected to be the consequence of tumor heterogeneity and the reason for real therapeutic problems. As far as a tumor contains a single, hormone-dependent cell population, androgen deprivation will result in an excellent clinical response with shrinking of the tumor and release of pain. A tumor composed of hormone-dependent and -independent parts will, upon androgen deprivation, initially also react with a good clinical response due to the shrinking of the hormone-dependent part of the tumor. However, elimination of the hormone-dependent part of the tumor may have the adverse effect of the hormone-independent part of the tumor, now without competition from hormone-dependent cell populations, gains advantages in growth. The predicted result is more rapid progression to disseminated disease.

Summary

Quantitative measurements of nuclear DNA by flow cytometry have been used in the investigation of bladder and prostate carcinomas. The subdivision of tumors into diploid, tetraploid and aneuploid enables further characterization of morphologically defined tumors.

Distinct differences in the distributions of ploidy of aneuploid grade 2 and grade 3 bladder carcinomas strongly suggest two different pathways of development of these gross chromosomally abnormal tumors. Significantly lower S-phase values of grade 2 tumors compared with grade 3 tumors of the same degree of ploidy enable the differentiation of these two tumor types. The proportion of S-phase cells is significantly related with the stage of invasiveness of bladder carcinomas.

In prostate carcinomas, the exponential decrease of diploid tumor with tumor stage, the increase of tetraploid tumors to maximum at intermediate stages, and the exponential increase of aneuploid tumors with stages after some delay strongly support the concept of the continuous development of the prostate carcinoma according to a three-compartment system. Further support for this concept is gained by the study of repeated fine-needle aspirates of the tumor and the observation of coexisting diploid and aneuploid cell lines in the developing tumor. An adverse effect of hormone deprivation on composed tumors by elimination of androgen-dependent and outgrowth of androgen-independent cell lines cannot be excluded.

References

Adolfsson J, Tribukait B (1990) Evaluation of tumor progression by repeated fine needle biopsies in prostete adenocarcinoma: modal deoxyribonucleic acid value and cytological differentiation. J Urol 144:1408

Adolfsson J, Rönström L, Hedlund P-O, Löwhagen T, Carstensen J, Tribukait B (1990) The prognostic value of modal deoxyribonucleic acid in low grade, low stage untreated prostate cancer. J Urol 144:1404

Batemen H (1910) The solution of a system of differential equation occurring in the theory of radioactive transformations. Cambridge Phil Soc Proc 15:423

Esposti P (1971) Cytologic malignancy grading of prostatic carcinoma by transrectal aspiration biopsy: a five-year follow-up of 469 hormone-treated patients. Scand J Urol Nephrol 5:199

Franzén S, Giertz G, Zajicek J (1960) Cytological diagnosis of prostatic tumors by transrectal aspiration biopsy: a preliminary report. J Urol 32:193

Harmer MH (1978) TNM classification of malignant tumours. UICC, Geneva

Hermanek P, Sobin L (1987) TNM classification of malignant tumours. UICC/ Springer, Berlin Heidelberg New York

Koss L (1979) Mapping of the urinary bladder: its impact on the concepts of bladder cancer. Hum Pathol 10:533

Mostofi FK, Sobin LH, Tovloni H (1973) International histological classification of tumours. WHO, Geneva

Tribukait B (1987) Flow cytometry in assessing the clinical aggressiveness of genito-urinary neoplasms. World J Urol 5:108

Wang N, Stenkvist B, Tribukait B (1992) Morphometry of nuclei of the normal and malignant prostate in relation to DNA ploidy. Anal Quant Cytol 14:210

Cytodiagnosis of Viral Infections

D.V. Coleman

Department of Cytopathology, St. Mary's Hospital, London W2, UK

Viruses are infectious particles that are composed of a nucleic acid core (either DNA or RNA) surrounded by a protein coat or capsid. The capsid is composed of distinct morphological units called capsomeres which determine the symmetry of the virus. Thus viruses may have helical symmetry if the capsid has a coil-like structure or an icosohedral symmetry. Some viruses, e.g., the herpes viruses, have an outer envelope which may be derived from the host cell membranes.

Viruses are obligate intracellular parasites which are species specific. They may infect a cell without apparently interfering with cellular structure or function. The viral genome may be present in the cell but replication does not occur except during periods of reactivation of the virus. This period of quiescence is termed latent infection. Reactivation is associated with virus replication and virus shedding. Reactivation too may be asymptomatic but more often is associated with cellular damage which usually leads to cell death. In its most destructive form virus replication may result in cell lysis or gross enlargement of the cell. Alternatively, it may result in the formation or intranuclear or intracytoplasmic inclusions or lead to cell fusion and the formation of multinucleate cells.

The cellular changes induced by viruses are termed the cytopathic effect (CPE) of the virus. The CPE is often very characteristic for a specific virus type or genus and may be regarded by pathologists as pathognomonic for a particular virus. Thus the presence of viral CPE in cytological or histological material is of diagnostic interest. However, it is important to remember that the morphological changes in the cell seen in the light microscope are only indirect evidence of the presence of viral infection. Definitive diagnosis depends on demonstration of virus particles, viral proteins or viral DNA in the cells. This can be achieved by virus isolation in cell culture, by immunocytochemical staining for viral antigens or by in situ hybridisation.

The cytopathic effect of viruses may be detected in cytological specimens (Coleman 1982). The large inclusion bearing cells characteristic of cyto-

Recent Results in Cancer Research, Vol. 133

megalovirus infection and the multinucleate giant cells of herpes simplex and herpes zoster were among the first to be recognised. More recently, changes due to the papovaviruses [human papillomavirus (HPV) and human polyomavirus] and molluscum contagiosum were added to the list. The cytological diagnosis of herpes simplex virus and HPV in cervical epithelium has been of particular interest to cytopathologists because of the association of these viruses with cervical cancer. Moreover, the increase in the number of patients whose immune system is impaired by drugs or disease has increased the frequency with which these viral changes may be found, and it is important for the cytopathologist to be aware of the range of cellular changes associated with these micro-organisms in view of their clinical significance.

The cytological manifestation of a number of common virus infections and some clinical implications of infection are discussed in this paper.

Herpes Simplex Virus

Herpes simplex virus (HSV) is a member of the herpes group of viruses which includes herpes simplex type 1, herpes simplex type 2, cytomegalovirus, Epstein-Barr virus, varicella/zoster virus, and human herpes virus 6. The herpes viruses are relatively large viruses, 125 nm in diameter, with a cuboidal symmetry and a DNA core. The capsid is surrounded by a lipid coat derived from the host cell membrane.

The HSV types 1 and 2 are antigenically distinct and infect the majority of the world's population early in life. After the primary infection, the viruses persist in latent form but reactivate at intervals, shedding virus particles. HSV type 1 classically infects the conjunctiva, oropharyngeal mucosa and the central nervous system. In immunosuppressed patients, HSV type 1 may cause acute oesophagitis or tracheitis. HSV type 2 is found most frequently in the genital tract. Neonatal infection can occur during parturition. The risk of potentially fatal herpes infection of the neonate is most likely to follow primary infection of the mother in the first trimester of pregnancy.

Infection of the skin and mucus membranes by HSV types 1 and 2 is characterised clinically by the formation of small vesicles in the epithelial layers. The vesicles contain an inflammatory exudate in which multinucleate giant cells may be found. The vesicles are often painful and may ulcerate and become secondarily infected. Both primary and secondary infections may be accompanied by fever and malaise.

The cytopathic changes of HSV infection can be found most commonly in cervical smears and in sputum samples and oesophageal brushings. They are characterised by the presence of giant cells formed as a result of fusion of the epithelial cells infected by the virus (Fig. 1). The giant cells may contain up to 50 nuclei. The nuclei usually appear glassy and structureless; mar-

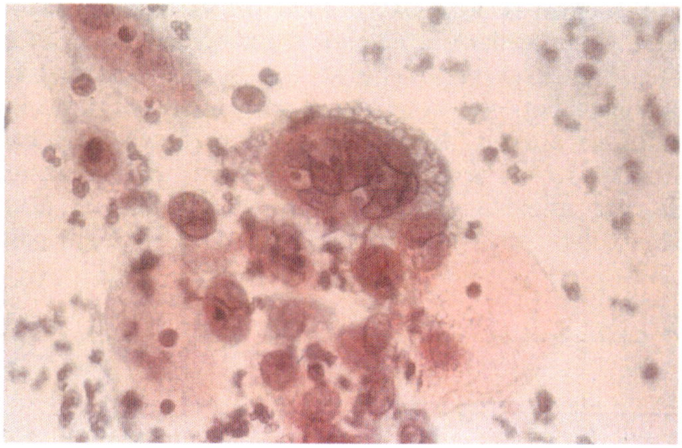

Fig. 1. Herpes infected cell in cervical smear showing multinucleation, ground glass appearance and moulding of nuclei. Papanicolaou stain; ×400

gination of the chromating and nuclear moulding are prominent features. The ground glass appearance of the nuclei reflects the fact that the chromatin is replaced by viral proteins. As the protein contracts, a typical Cowdray type A inclusion may be formed. Unless the cytopathologist is aware of the bizarre appearances produced by the herpes viruses, these virus-infected cells may be misinterpreted as malignant cells or even, in cervical smears, as trophoblast. Confirmation of the viral nature of the cells can be obtained by reprocessing the cells for electron microscopy and visualising the virus particles.

An association between HSV type 2 and cervical cancer was noted 20 years ago by Naib et al. (1966) who observed viral CPE in cervical smears from women with cervical neoplasia. This led to speculation that HSV type 2 plays a role in cervical carcinogenesis. Subsequent epidemiological studies based on serological surveys of cases and controls and cross-sectional prospective studies of a defined female population were unable to demonstrate a role for the virus in cervical carcinogenesis. Moreover, HSV DNA could not be consistently found in the cervical neoplastic lesions.

The cytopathic effect of the varicella/zoster virus is indistinguishable from those induced by the HSV. In practice, the specific virus type causing the morphological changes is deduced from the clinical presentation of the herpetic lesions. For example, multinucleate giant cells with intranuclear inclusions in a cervical smear are unlikely to be due to zoster. On the other hand, if they are found in vesicles which have a neurotropic distribution, zoster will need to be eliminated.

On occasion, the vesicles due to herpes infection may be difficult to distinguish from those due to pemphigus vulgaris. Confusion between the two diseases is most likely when the vesicles or bullae are found in the mouth – particularly as ulceration and secondary infection is frequent at that site. Tzanck (1948) described a cytological test that can be used to discriminate between these two diseases. Scrapings taken from the vesicles or ulcers are very informative. The presence of multinucleate giant cells is indicative of herpes infection. Pemphigus is characterised by numerous discrete rounded epithelial cells with prominent nucleoli reflecting the acantholytic process which is characteristic of pemphigus.

Cytomegalovirus

These viruses are members of the herpes group and at the electron-microscopic level are indistinguishable from herpes zoster or herpes simplex. However, the behaviour of these viruses in tissue culture is quite distinct as they grow slowly and have a very restricted growth range. Primary infection may occur at any time and is frequently related to the onset of sexual activity. In adolescents and young adults, primary infection is frequently associated with a mononucleosis-like illness or viral pneumonia. Many primary infections are asymptomatic and the virus remains in a latent state until reactivated. Reactivation occurs in allograft recipients, acquired immunodeficieny patients or other patients who are severely immunocompromised. The sites most commonly affected are the endothelial cells in lung and kidney. The CPE is characterised by the presence of enlarged cells containing eosinophillic intranuclear inclusions (Fig. 2). These are most frequently seen in urine sediment but may also be found in bronchoalveolar lavage specimens or fine-needle aspirates from thyroid or salivary gland.

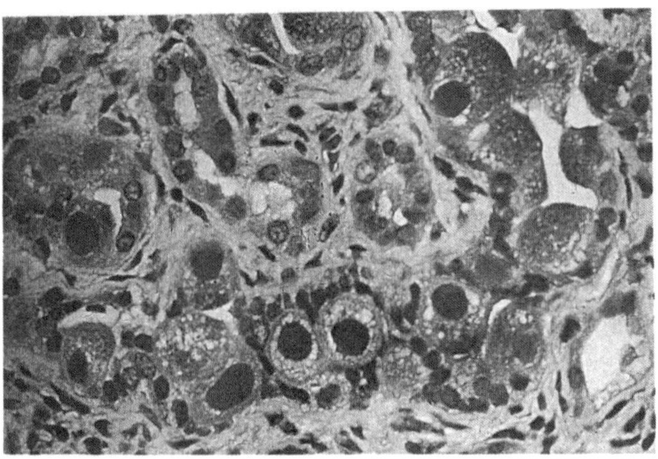

Fig. 2. Cytomegalic inclusions in endothelial cells in biopsy of renal cortex from renal allograft recipients. Haematoxylin and eosin; ×400

The large intranuclear inclusions confer a "bird's eye" appearance to the cell. Numerous small cytoplasmic inclusions may also be seen.

Cytomegalovirus infection rarely may be transmitted in utero from mother to fetus. The large inclusion bearing cells have been described in urine of neonates who have acquired infection in this way.

Human Polyomavirus

The human polyomaviruses are small viruses 42–45 nm in diameter with a DNA core surrounded by a capsid composed of 72 capsomeres arranged in icosahedral symmetry. Several antigenically distinct polyomavirus types have been identified, of which the most common are BK virus and JC virus. Serological studies have shown that primary infection is common in childhood. It is usually asymptomatic and followed by latent infection. In the majority of individuals with a normal immune response, the virus remains in a quiescent state throughout life. Reactivation of latent virus occurs in individuals whose immune reaction is impaired. When this occurs, replicating human polyomavirus particles (both BK and JC) can be found in the urothelium. JC virus also has a predilection for the oligodendroglial cells of the brain. Thus reactivation of JC virus is associated with the onset of progressive multifocal leucoencephalopathy – a rare but invariably fatal demyelinating disease of brain. Reactivation of BK virus is often symptomless although it may be associated with urinary obstruction.

Diagnosis of human polyomavirus infection is difficult. Virus isolation techniques are tedious and definitive diagnosis depends on in situ hybridisation or immunocytochemical studies. Thus cytology can be a very useful tool for the diagnosis of this infection. Replication of virus is intranuclear and is associated at the light-microscopic level with the formation of large basophilic intranuclear inclusions (Figs. 3, 4). The inclusions are formed almost entirely of virus particles arranged in pseudocrystalline array. The inclusion-bearing cells may be found in smears of urine sediment from renal allograft recipients. They have also been demonstrated in urine from pregnant women and males with prostatic obstruction. These findings suggest that obstruction to urinary flow may contribute to virus reactivation.

These inclusion-bearing urothelial cells with their large intact basophilic inclusion have in the past been mistaken for malignant transitional epithelial cells. They can be distinguished from the latter by virtue of the thickening of the nuclear membrane and the homogeneous glassy appearance of the inclusion body. In some urothelial cells, the inclusion bodies may appear to be fragmented and poorly preserved. Only the thickened nuclear membrane remains intact.

Although the polyomaviruses appear to have an oncogenic potential in vitro, the presence of these viruses in the urinary tract has not been associated with the development of urothelial cancers.

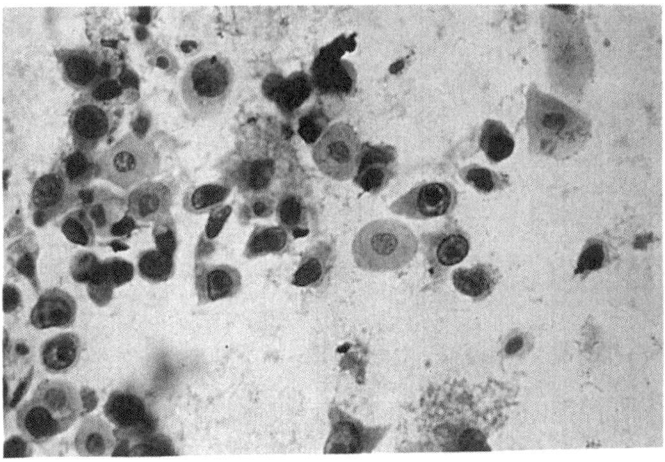

Fig. 3. Basophilic intranuclear inclusions due to BK virus in transitional cells in smear of urinary sediment from renal allograft recipient. Papanicolaou stain; ×400

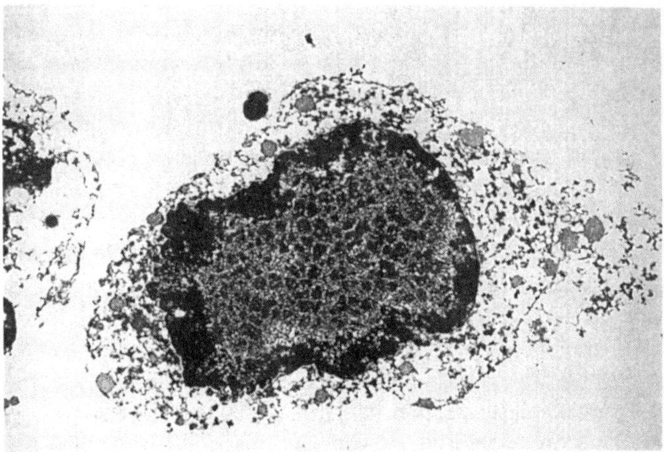

Fig. 4. Electron micrograph of polyomavirus-infected transitional cell showing intranuclear virus particles in pseudocrystalline array. Note thickened nuclear membrane, also evident in Fig. 3. TEM uranyl acetate; ×1000

Human Papillomaviruses

The papillomaviruses are small DNA viruses of 52–55 nm which have a morphology very similar to that of polyomaviruses (Fig. 5). They cause papillomas, warts or other benign epithelial lesions in a wide range of mammals and birds. Each member of the group is species specific. For many

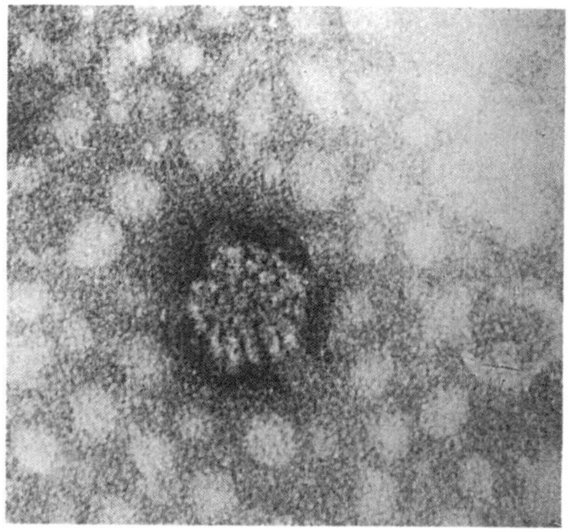

Fig. 5. Papillomavirus particle of 55-nm diameter – note slightly skewed icosahedral symmetry. Negative staining of scrapings from common skin wart; phosphotungstic acid

years little was known about biological properties of the papillomaviruses as suitable methods for propagating the viruses in tissue culture did not exist. With recent developments in molecular biological technology and immunocytochemistry, our knowledge of the properties of this group of viruses has increased (Ward and Coleman 1992).

There are over 50 types of HPV which can be distinguished on the basis of the homology between their genomes as determined by cross-hybridisation between the separated DNA strands. The DNA of each virus type has less than 50% homology with any other virus type. Each HPV type is associated with a characteristic clinical lesion. For example, types 1 and 4 are associated with plantar warts. Types 6, 11, 16, 18, 33, 34, 55 and 57 are associated with genital lesions.

Human papillomaviruses are epitheliotrophic. The viral DNA can be found in the basal cells of the epithelium but the whole virion is found only in the uppermost cell layer in the terminally differentiated keratinocyte. This requirement for keratinisation for the purposes of virus replication has prevented the development of a successful tissue culture system. The viral DNA is found in an episomal form or integrated form – the latter form being most common in malignant tissue.

Clinical Features

Clinically HPV infection manifests itself by the presence of exophytic warts or papillomata. The most common lesions are cutaneous and occur on the hands and plantar region of the foot. Warts and papillomata are also to be found on the mucous membranes of the genitalia where they are termed condylomata accuminata. The viruses are also the causative agents of juvenile laryngeal papillomatosis.

Recent studies using colposcopy have revealed that HPV infection in the genital tract may also be present in a subclinical form. These lesions can be demonstrated by the application of 5% acetic acid to the mucosa of the cervix. Because these lesions lack the papillary appearance of exophytic warts, they have been termed non-condylomatous warts, flat warts or subclinical papillomavirus infection.

Histological Diagnosis

Cutaneous warts are characterised by papillomatosis, hyperkeratosis, hypertrophy of the basal layers and some degree of proliferation of the intermediate layers (acanthosis). Within the superficial layers of the epithelium, koilocytes may be seen, together with the occasional multinucleate cells and individual highly keratinised squames (dyskeratotic cells) (Fig. 6). Thus, apart from the koilocytes, the changes are not specific for papillomavirus infection. Even the identification of koilocytes, which are characterised by

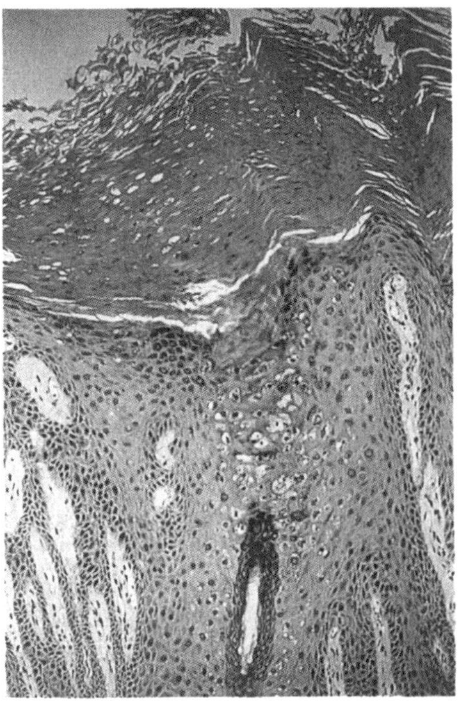

Fig. 6. Histological section of common skin wart showing papillomatosis, acanthosis, hyperkeratosis and koilocytosis. Haematoxylin and eosin; ×100

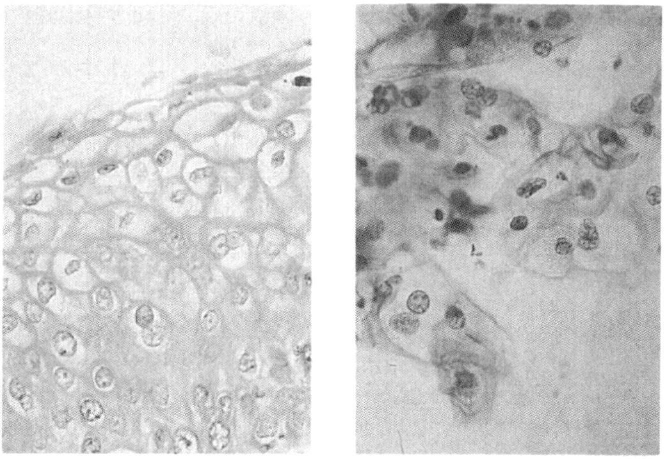

Fig. 7. Histological and cytological appearances of HPV infection of cervix. Both biopsy and cervical smear shows koilocytotic changes. Biopsy stained with haematoxylin and eosin; ×100. Cervical smear stained with Papanicolaou stain; ×400

the large area of perinuclear clearing and thickened rim of cytoplasm, is a subjective exercise.

Genital warts (condylomata accuminata) have a similar histological appearance to cutaneous warts except that keratinisation of the surface layers may be less marked. Subclinical lesions in the genital tract (flat warts, non-condylomatous infection) do not exhibit the papillomatosis which is a feature of exophytic warts (Fig. 7).

Diagnostic Methods

The definitive diagnosis of HPV infection depends on demonstration of the virus or its antigens by electron microscopy, immunocytochemical staining and by DNA hybridisation techniques. Each method has its particular limitations.

Electron microscopy and immunocytochemical staining for common wart virus antigen can be used to confirm the presence of HPV in biopsy material. Although these techniques are highly specific, they lack sensitivity and nowadays are rarely used.

At present the only means of accurately determining the specific HPV type is to *hybridise the DNA* extracted from the tissue under investigation using virus specific probes. The hybridisation techniques in current use include Southern blotting, filter in situ hybridisation (FISH), dot blot/slot blot techniques, polymerase chain reaction (PCR) and in situ hybridisation.

Table 1. Prevalence of HPV in normal female genital tract

Reference	Population studied	Method	Prevalence HPV DNA (%)
De Villiers et al. (1987)	10 000 healthy women	FISH	12
Wickenden et al. (1987)	103 healthy women	Dot blot	23
Schneider et al. (1988)	92 pregnant women	Southern blot	28
Tidy et al. (1989)	110 healthy women	PCR	70

The techniques vary in their sensitivity and specificity. Southern blotting gives highly specific results but requires a large amount of DNA and is best suited to biopsy material, whereas FISH and dot blot/slot blot techniques require less DNA. Unfortunately, interpretation of the signals produced by the FISH and dot blot assay is subjective. PCR is very sensitive and theoretically can detect a single virus genome in 5 mg DNA but is subject to problems of contamination. Thus stringent controls are necessary if the results of PCR are to be accepted as valid. Recent studies of the epidemiology of HPV 16 in women with normal cervical smears emphasises the diversity of results obtained with the various hybridisation techniques. The prevalence ranges from 12% using Southern blot to 70% using PCR (Table 1).

Cytological Diagnosis

The changes seen in histological section are reflected in cervical smears. The only cell that serves to characterise HPV infection is the koilocyte, although anucleate keratinised squames, dyskeratotic cells and multinucleate cells may be seen in the smears. The nucleus of the koilocytes may show a slight nuclear enlargement which does not amount to dyskaryosis. Studies to compare the sensitivity of cytology with DNA hybridisation have shown that both techniques have a margin of error of 10% or more (Table 2).

Table 2. Correlation between HPV DNA and cytological features in the total group (n = 164)

	HPV DNA			
	Present		Absent	
	(n)	(%)	(n)	(%)
Cytological features present	47	29	19	12
Cytological features absent	17	10	81	49

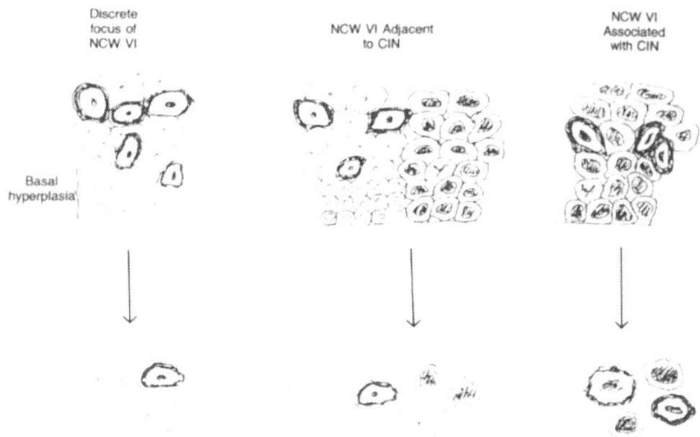

Fig. 8. Histological and cytological presentation of HPV infection in the cervix. *Left* to *right*; HPV infection in absence of CIN; HPV infection adjacent to CIN; HPV infection superimposed on CIN

HPV Infection and Cervical Cancer

Histological and cytological studies have shown that the morphological changes associated with HPV infection can be detected in intraepithelial neoplastic lesions and invasive squamous cancers arising in the genital tract. The presence of the viruses in these lesions is manifest by the appearance of koilocytes in the neoplastic tissue. The koilocytes themselves may have a bizarre nuclear structure indicating that the virus may be replicating in the tumour cells (Fig. 8).

The hypothesis that HPV have a role to play in the development of the carcinomas stems from the observation that in the cervix, HPV 16 DNA can be detected in a large proportion of high-grade squamous intraepithelial lesions and invasive squamous cancer, whereas HPV 6 and 11 are found most commonly in condylomata and in low grade squamous intraepithelial lesions. However there is still uncertainty as to whether the association between HPV 16 and these high-grade lesions is causal or casual.

Evidence to support a causal role is derived mainly from studies of the viral genome. The genome of HPV 16 is a supercoiled double-stranded molecule of DNA which consists of about 8000 base pairs. The early coding region of the viral genome contains seven open reading frames (ORFs) which code for proteins which control virus transcription, transformation and replication. In vitro studies have shown that the protein coded for by the E6 ORF of HPV 16 binds to p53 protein which is a protein found in normal cells which controls cell proliferation. Thus, through production of

E6 protein, HPV appear to be able to control cell proliferation and increase the mitotic activity of the epithelium.

Although this is compelling evidence for an oncogenic role for HPV 16 in the cervix, other evidence in favour of a casual role cannot be ignored. HPV 16 has not been demonstrated in all cervical cancers and is rarely found in squamous cancers arising in sites other than the uterine cervix, i.e., in the vulva, vagina or penis. In fact in the United States the HPV types most commonly found in cervical cancers are HPV 18 and 31. Moreover, HPV 16 is not uncommon in the normal woman and its presence does not appear to be associated with increased cancer risk. A plausible explanation for these findings is that cervical cancer may arise through a variety of oncogenic mechanisms and HPV 16 infection is but one of them.

Current research is now directed at identifying alternative carcinogenic pathways which can account for these diverse findings.

References

Coleman DV (1982) Cytological diagnosis of virus infected cells in cervical cancers: value in gynaecologic and obstetric practice. Diagn Gynaecol Obstet 4(4):363–374

De Villiers EM, Wagner D, Schneider A, Wesch H, Miklaw H, Wahrendorf J, Papendick U, Zur Hausen H (1987) Human papillomavirus infections in women with and without abnormal cervical cytology. Lancet II:703–706

Naib ZM, Nahmias AJ, and Josey WE (1966) Cytology and histopathology of cervical herpes simplex infection. Cancer 19:1026–1031

Schneider A, Hotz M, Gissman L (1987) Prevalence of genital HPV infection in pregnant women. Int J Cancer 198–201

Tidy JA, Parry GCN, Ward P, Coleman DV, Peto J, Malcolm ADB, Farrell PJ (1989) High rate of HPV 16 infection in the female population. Lancet 298:14–18

Tzanck A (1948) Le cytodiagnostic immèdiat en dermatologie Ann Dermatol Syphiligraph 8:205–218

Ward P, Coleman DV (1992) Papovavirus. In: Greenward D, Slack R, Reuther J (eds) Medical microbiology, Chap 46. Churchill Livingstone, Edinburgh, pp 519–526

Wickenden C, Malcolm A, Byrne M, Smith C, Anderson C, Coleman DV (1987) Prevalence of HPV DNA and viral copy numbers in cervical scrapes from women with normal and abnormal cervices. J Pathol 153:127–135

Cytology of Viral Infections in Gynecologic Smears

D. Wagner

Cytologisches Laboratorium, Evang. Diakoniekrankenhaus, Burgunderstr. 1, 79104 Freiburg, Fed. Rep. of Germany

Human papilloma viruses (HPV) produce infections of the female anogenital tract and are associated with precancerous and cancerous lesions of the vulva, vagina, and uterine cervix. HPV infection is considered to be an unequivocal causative factor for the development of cervical cancer.

Clinical HPV infections (i.e., condylomata acuminata) can be seen with the naked eye and are diagnosed macroscopically. *Subclinical HPV infections* can be discovered by colposcopy, cytology, and histology. *Latent HPV infections* can only be detected by molecular-biological methods such as DNA hybridization techniques or, under special conditions, by serology. These forms of infection can change and may show different manifestations at each examination.

Most, if not all, precancerous and cancerous lesions of the vulva, vagina, and uterine cervix are necessarily preceded by an HPV infection occurring at any time in the past life of the patient. Additional factors must come into play to transform the infected epithelium to a precancerous lesion and finally to cancer.

Since we do not know those factors at the present time, we still rely in clinical practice on the performance-proven morphological methods of early cancer detection: colposcopy, cytology, and histology. HPV DNA hybridization techniques can be used as an additional method to verify a morphologically suspicious lesion whether or not associated with an HPV infection, making them helpful for quality control of morphology.

HPV typing, however, does not seem to be of clinical significance since it is not relevant for therapy. Recent follow-up studies have clearly shown that oncogenic and nononcogenic HPV types may vary considerably in the same patient when examined several times.

When using hybridization techniques for clinical purposes, one should demand high specificity and reasonable sensitivity of the method chosen. The application of HPV hybridization techniques should be restricted to highly experienced laboratories which have different methods available that are suited to the special problem of clinical concern.

Recent Results in Cancer Research, Vol. 133

Immunocytochemistry in Diagnostic Cytology

P. Dalquen, G. Sauter, R. Epper, B. Kleiber, G. Feichter, and F. Gudat

Institute for Pathology, Division of Cytology, University of Basel, CH-4003 Basel, Switzerland

Introduction

Immunocytochemistry is now a well-established tool for tumor typing in surgical pathology. However, in diagnostic cytology it has not yet been generally accepted. Koss (1990) says that "... the staining of cytologic preparations with a battery of antibodies to the various cell components or products is very costly and rarely rewarding, ... the results of immunocytochemistry vary according to the batch of antibodies and the technical skills of the laboratories, the interpretation of the results is not always easy, and the problem with borderline positive stains is often perplexing." Other authors (Chess and Hajdu 1986; van Flens et al. 1990; Mason and Bedrossian 1986), too, are not as enthusiastic as some of the pioneering cytolgists (Nadij 1980). Chess and Hajdu (1986) found only 63% positive reactions and 88% negative reactions to be in agreement with the cytologic diagnosis, Van Flens et al. (1990) estimated that immunocytochemistry contributed to diagnosis in about 50% of cases.

There are several reasons for the different acceptance of immunocytochemical methods by the two morphological specialities:

1. Compared to histology the number of cytologic preparations is usually limited so that only few incubations can be performed.
2. Even if several smears are available in a given case, it is not known which of the smears contains the diagnostic cells until the smears have been stained for conventional light-microscopic evaluation.
3. Since the cells in cytologic preparations retain their spherical shape, cytoplasmic reaction products may mask the chromatin structure of the nuclei which is so important for the differentiation between malignant and nonmalignant cells.
4. The superimposition of cells in cytologic preparations makes reading of the immunocytologic reactions difficult.

Recent Results in Cancer Research, Vol. 133

5. The provenance of cytologic specimens from different sites calls for special preparatory methods because of different content of cells and proteins.
6. The quality of specimens varies considerably with the technical skill of both the laboratory and the clinician who has sent the specimens.

The purpose of this paper is to analyze the significance of immunocytochemistry in cytology on the basis of our own 8-year experience. It will show how to make the most of immunocytochemistry with the limited cellular material received from clinicians.

Material

Between 1985 and 1992 immunocytochemical examinations were performed on 507 out of 70000 cytologic specimens (0.7%); 87.4% of the specimens examined were effusions and fine-needle aspirates. Beyond that a small number of specimens prepared from sputum, bronchial secretion, brush biopsies, and cerebrospinal fluid were studied (Table 1).

Table 1. Specimens used for immunocytochemical incubations

Specimen	*n*
Sputum/bronchial secretion	23
Effusion	215
Brush biopsy	13
Fine-needle aspiration	205
Cerebrospinal fluid	25
Tissue medium	9
	490

Table 2. Indication for immunocytochemistry on cytologic specimens

Indication		*n*
Tumor typing		274
Small cell tumor	44	
Large cell tumor	25	
Other tumors	205	
Mesothelioma		65
Other indications		4
Tumor/reactive lesion		147
		490

The immunocytochemical examination was indicated in 274 cases (55.9%) for tumor typing; in 65 cases to differentiate mesothelioma from adenocarcinoma; and in 147 cases, mainly effusions, for the differentiation between benign (reactive) and malignant lesions (Table 2).

Methods

Fixation

Smears of fine-needle aspiration and brush biopsies were usually fixed with commercially available *spray* containing polyethyleneglycol in ethanol. All other smears prepared by the cytologic laboratory were fixed with Delaunay's solution (100% ethanol and 100% acetone in equal parts mixed with 0.5 ml/ 1000 1 *M* trichloric acetic acid).

Cytologic Staining Method

The smears were usually stained with *Papanicolaou's stain*. The results of immunocytochemistry do not depend on the staining method (Li et al. 1987; Yam 1990). Only minor modifications to the immunocytochemical methods (e.g., change of antibody concentration) may be necessary for each stain. The advantage of Papanicolaou's stain in immunocytochemistry is wet fixation, which guarantees preservation of both the nuclear structure and the immunoreactivity even after the smears have been stored for a long time. It can easily be combined with immune stains. The advantage of the *May-Grünwald-Giemsa* (MGG) *stain* is the clear representation of cytoplasmic details (Yam 1990). However, the immunoreactivity of the air-dried cells of uncovered MGG-stained smears disappears at room temperature within a few days (Dalquen et al. 1986; Dinges et al. 1989). If immunocytochemical incubations are performed within 2–3 days, excellent reactions can be achieved. *Methyl green* has been recommended as a counterstain since it stains only nuclei. The nuclear staining can be removed by a green filter when black and white photographs are taken for publication (Osamura et al. 1986).

Storage of Specimens

Unstained smears can be kept for several hours in Delaunay's solution or in xylene without evident loss of immunoreactivity. This is only advisable if all slides contain enough diagnostic cells. Usually the diagnostic cells are unevenly distributed throughout the slides. Therefore, the slides were first stained and covered with Eukitt (O. Kindler GmbH & Co. D-7800 Freiburg) and cover glass before selecting the specimens for immunocytochemistry. In covered preparations immunoreactivity of many epitopes is preserved for many years. If immunocytochemistry was indicated the covers were removed with xylene (see below).

For many antibodies against cluster-defined differentiation antigens of leukcocytes (CD antigens) it is necessary to store air-dried and subsequently acetone-fixed smears (cytospin preparations) welded in plastic foil at −70°C. This step of lyophilization and fixation improves the antibody reaction apparently by uncovering the corresponding epitopes. The smears retain their reactivity for many months.

Cellular material can be preserved for prospective studies by the same storage procedures.

Special Preparatory Techniques

Sometimes concentration by centrifugation and other methods developed to study sparse cellular material from routine cytology were applied.

The most simple method is to encircle several fields on a smear using a Pap Pen (Daido Sangyo Co., Ltd. Tokyo, Japan) for subsequent incubation with different antibodies. This method is only applicable if enough diagnostic cells are evenly distributed over the slide.

Cytospin centrifugation is in our opinion the most important method of cell concentration in body fluids containing only few cells, such as urine or cerebrospinal fluid. However, it may also be applied to fine-needle aspirates. The whole aspirated material or cells left in the hub of the needle after preparation of the usual Papanicolaou smears and normally discarded are flushed into 3 ml tissue culture fluid and then centrifuged (Brown et al. 1989; Kung et al. 1990). Using a 23-gauge needle, sufficient material is usually aspirated for up to ten or even more cytospin preparations containing a constant number of cells. We found, however, that the cell yield is often higher in aspirates from lymphomas than from epithelial tumors. The preparations can be differently fixed so that optimal conditions for different antibody incubations are available. Cell loss caused by centrifugation is compensated by utilization of the whole aspirated material.

The *cell block technique* has rarely been used. Cells suspended in a fluid medium are concentrated by centrifugation, the pellet fixed with formalin, embedded in paraffin and cut in series. This technique, first introduced at the end of the last century, still enjoys great popularity, especially among pathologists, for several reasons:

1. Interpretation of the results is easier and less time consuming as the section area is small and some histologic patterns such as tubuli are preserved.
2. Up to 20 incubations can be performed on one sample.
3. Loss of the standard smear for the conventional cytologic diagnosis is avoided.

4. The paraffin blocks are easy to store (Kung et al. 1990).

Several major disadvantages, however, stand against it:

1. The main disadvantage is that this technique is only applicable to material containing a sufficient number of the relevant cells. If an effusion contains only a small portion of diagnostic cells, the chance of detecting them in the sections of the cell block is poor, whereas they would not be missed in the Papanicolaou-stained smear.
2. The nuclear details are not as brilliantly preserved as in the acetone- or alcohol-fixed Papanicolaou smears.
3. The preparation of a histologic slide of cytologic material is more time consuming than making smears.
4. The staining intensity is usually stronger in smears than in cell blocks (Mason and Bedrossian 1986).

Imprints from fresh tissues have been used in selected cases, especially as a complement to frozen sections during surgery. The technique is most useful in the differential diagnosis of neoplastic and non-neoplastic lymphomas (Banks et al. 1983) and other small cell tumors.

Antibody Panel

Most antibodies used in our laboratory on routine Papanicolaou-stained cytologic specimens are commercially available. Specifity and optimal antibody concentration were first tested on paraffin sections of formalin-fixed tissue or on frozen sections of fresh tissue. The following general rules could be derived from pretesting:

1. Antibodies not working on paraffin sections give generally no specific staining on cytologic smears which are fixed wet as described above.
2. On cytologic material an antibody concentration twice as high as for histologic sections has to be chosen. There are two possible reasons for this: (a) in wet-fixed smears the cells are intact and the cellular membranes tight so that the access of the antibody to cytoplasmic epitopes is hampered; (b) the epitopes are partly damaged by the slightly acid solutions of the preceding Papanicolaou stain.
3. Intermediate filaments, cytoplasmic or membrane-bound glycoproteins, and cytoplasmatic enzymes can be studied on cytologic preparations fixed wet with alcohol, acetone, or Delaunay's solution. Many CD antigens of leukocytes, however, can only be demonstrated on air-dried and subsequently acetone-fixed preparations. The same is valid for cell-bound immunoglobulins in serous fluids, where the cells are coated by immunoglobulins of the blood plasma.
4. The use of monoclonal antibodies is to be preferred because contamination with unrelated antibodies is avoided. However, loss of the relevant epitope will lead to false-negative results. If this does not happen,

monoclonal antibodies work better than polyclonal antibodies on cytologic material.

There were six main indications for immunocytochemistry:

1. Malignant versus reactive cells.
2. Small cell carcinoma versus lymphoma.
3. Large cell carcinoma versus lymphoma or melanoma.
4. Carcinoma versus mesothelioma.
5. Spindle cell tumors, for example leiomysarcoma, versus neurogenic tumors or cells from organizing tissue.
6. Differentiation of special tumor types such as metastases of prostatic carcinoma, ovary carcinomas, carcinoids, or other neuroendocrine tumors.

Immunocytochemistry was performed to confirm malignancy (indication 1), if (a) atypias were found in effusions by light-microscopic screening; (b) if normal and neoplastic lymphocytes had to be differentiated in purely lympocytic effusions; and (c) if low-malignant lymphoma was suspected in fine-needle aspirates from lymph nodes, but reactive follicular hyperplasia could not be precluded.

Which antibody combination was used depended on the indication (Table 3). The first five indications were answered by well-aimed application of two or three out of a panel of maximally 10–15 antibodies. For a more detailed diagnosis of tumor type (indication 6), however, antibodies were selected from a large battery used at our institution in histology, too.

All the antibodies listed in Table 4 work on cytologic specimens fixed with ethanol, acetone, or mixtures of both (Delaunay's solution) and stained with Papanicolaou's stain. The epitopes of many CD markers of leukocytes,

Table 3. Panel of antibodies for immuncytochemical examinations in diagnostic cytology

Indication	Antibody
Tumor/reactive	
Effusion with atypias	MAb Lu-5, mAK HEA 125, PAB CEA, mAB BerEP4
Lymphocytic effusion	MAb UCHL-1, mAB L26
FNAB from lymphoma	MAb UCHL-1
Mesothelioma/carcinoma	MAb HEA 125 or MAb BerEP4, PAb CEA,
Tumor type	MAb Lu-5, MAb Vim, MAb NSE, MAb Des, MAb PSA, MAb SPP MAb UCHL-1, MAb L26 etc.
Small cell carcinoma/lymphoma	MAb Lu-5, MAb LCA
Large cell carcinoma/melanoma, etc.	MAb Lu-5, PAb S100, MAb HMB 45

Table 4. Antibodies most frequently used in this study

Antibody	Producer	Epitope
Lu-5[1]	BMA	Basic and acid cytokeratins ("pan-keratin")
Keratin[2]	Dakopatts	Keratins, predominantly of MW 56 and 64 kD
EMA[1]	Dakopatts	Membrane-bound antigen of MW 265–400 kD
BerEP4[1]	Dakopatts	Membrane-bound glycoproteins of MW 34 and 49 kD
HEA125[1]	Readysysteme AG	Cell surface glykcoproteid of MW 34 kD
CEA (A5B7)[1]	Dakopatts	Carcinoembryonic antigen
NSE[1]	Sanbio	Neuron-specific enolase
LCA[1]	Dakopatts	Leucocyte common antigen (CD45)
UCHL1[1]	Dakopatts	T lymphocyte specific component of CD45 (CD45R)
L26[1]	Dakopatts	B lymphocyte specific polypeptide of MW 33 kD (CD20)
HMB45[1]	Enzo Diagn. Inc.	Melanoma-specific cytoplasmic antigen
Anti-PSA[1]	Dakopatts	Human prostate-specific antigen, 33 kD
Anti-PAP[1]	Dakopatts	Human prostatic acid phosphatase
Anti-vimentin[1]	Boehringer	Vimentin

[1] Monoclonal antibody.
[2] Polyclonal antibody (PAb).
MW, molecular weight.

Table 5. Number of antibodies applied per case

Antibodies (n)	Cases (n)
1	103
2	206
3	115
4	33
5	22
6	8
7	1
8	2
	490

however, are too unstable and need an alternative preparatory technique (see below).

In more than 100 cases material for incubation was available with only onc antibody. On average, however, between two and three antibodies were applied per case (Table 5).

Incubation Methods

There are several immunocytochemical methods which have all been applied on cytologic material. However, some have considerable disadvantages in routine cytology.

We decided on the *avidin–biotin complex method* (ABC method) (Hsu et al. 1981) and the more stable substrate diaminobenzidine. With this method, background staining is minimal so that washing of cells from serous fluids is not necessary: the smears were first screened as usual. When immunocytochemistry was indicated, the cover slip was removed with xylene. Uncovering can be accelerated by using the more toxic methyl-ethyl-keton instead of xylene. Destaining of the smears with 1% periodic acid as suggested by some authors (Chess and Hajdu 1986) was found unnecessary. After rehydration and stopping of the endogene peroxydase with 1% H_2O_2 in methanol, the smears were incubated with the antibodies according to the ABC-Elite (Vector-Stain) method following the producer's instruction. This kit contains avidin DH and biotinylited horseradish-peroxidase H, which improves the formation of complexes during the immunoperoxydase reaction.

All antibodies were incubated overnight at 4°C in a moist chamber. Using the immunostaining center Sequenza (Shandon) 50% of antibody solution and about 30% of working time is saved. 3,3′Diaminobenzidin × 4HCI was used as a substrate for the peroxydase reaction. Between the various steps of the procedure smears were washed with PBS. The antibody concentration was regarded as optimal if a granular reaction product within the cells could clearly be seen and background staining was low or absent.

After the immunocytochemical procedure, the nuclear stain was briefly restored with Böhmer's hematoxylin if necessary. Surplus hematoxylin was washed off with 0.5% hydrochloric acid and the smears dyed blue with tap water. Thereafter they were gently restained, dehydrated, put in xylene and covered again with Eukitt and a cover slip.

Incubation at room temperature would call for observance of strict incubation time to avoid background staining. Beyond that, at room remperature the salt concentration of buffer solutions can be kept less constant so that cell damage by osmosis cannot be completely precluded. In case of high avidity between epitope and antibody (Lu-5!), however, the incubation time can be reduced to a few minutes so that no background staining occurs.

Immune peroxydase (peroxidase–anti-peroxidase – PAP, ABC method) and *immune alkaline phosphatase techniques* (immune alkaline phosphatase – IAP; alkaline phosphatase–anti-alkaline phosphatase – APAAP) are especially suitable to cytology as cytologic details of normal versus malignant cells remain clearly recognizable and the staining product is stable. To circumvent the problem of endogenous peroxidase, some authors prefer alkaline phosphatase techniques (To et al. 1983; Yam et al. 1987). But endogenous phosphatase may also be a problem in some tissues, e.g., in osteoblasts and intestinal cells. It is at least partly blocked by 1–5 m*M* levamisole (cimetidine) (Conway de Macario et al. 1986; Wadsley and Watt 1987). Endogeneous peroxidase is a minor problem in the peroxydase techniques since it is reliably blocked with hydrogen peroxide and methanol. The sensitivity of the methods may partly depend on the antibody used in the test (Tello and Johnston 1989).

Indirect immunofluorescence is a method sensitive enough for most antigens, especially if applied to fresh cellular material not altered by fixation or any staining procedure. All kinds of antigens can be demonstrated with this method (Altmannsberger 1988; Bruderman et al. 1990; van Croonen et al. 1988; Domagala et al. 1989a,b; Ramaekers et al. 1984; Walts and Said 1983). It is not suited for routine cytology for two reasons: (a) tumor cells cannot be clearly differentiated from normal cells as nuclear structure and other criteria of malignancy cannot be judged; (b) the reaction has to be documented photographically because of instability of fluorescence. Double labeling was not performed in this study. We reserve this powerful method for scientific questions.

Evaluation

The reaction was regarded as positive if the cells were decorated by a granular yellowish-brown reaction product. In each case the staining results were compared to the original cytologic findings and to the final clinical, bioptic, or autoptic diagnosis to decide whether they were *diagnostic, confirmative*, or *equivocal*. The results were regarded as diagnostic if the diagnosis could only be made by immunocytochemistry, as confimative if immunocytochemistry merely confirmed the diagnosis made on the usual Papanicolaou-stained smears or by previous biopsy.

The immunocytochemical reactions were mainly controlled by positive inner controls. Regrettably, negative controls (replacement of the specific antibody layer by PBS) were rarely possible bcause of the limited cellular material in the routine cytologic specimens. If at all possible, the immunoreactivity of the cells was controlled by at least one further antibody (Lu-5 or anti-vimentin) whose reactivity was beyond doubt.

Results

Differentiation of Reactive from Malignant Cells

In 147 specimens (mainly effusions) detection of tumor cells was tried with various antibodies. The results were equivocal in 27 cases (18.4%). The three main diagnostic problems were (a) differentiation of low-grade lymphoma from follicular hyperplasia in lymph node aspirates; (b) recognition of lymphoma in lymphocytic effusions; (c) differentiation of mesothelial cells from carcinoma cells in effusions.

Differentiation of Low-grade Lymphoma from Follicular Hyperplasia in Lymph-Node Aspirates

The differential diagnosis of reactive lymphadenitis with follicular hyperplasia versus centroblastic–centrocytic non-Hodgkin lymphoma may be extremely difficult, especially in Papanicolaou-stained smears (Dalquen 1986; Kline et al. 1984; Levitt et al. 1985; Lopez-Cardozo 1980; Zajicek 1974). The diagnosis can be improved by the use of a line-specific antibody. We stained wet-fixed smears from fine-needle aspirations of lymph nodes with the BAb UCHL1 (CD45R) which recognizes an epitope occurring in thymocytes and activated T lymphocytes but only in a small portion of resting T cells. The antibody reacts well on fixed tissues and Delaunay- or spray-fixed cytologic specimens.

The results of 63 examinations are given in Table 6. Fewer than 14% UCHL1-positive cells in a smear were found only in malignant B lymphoma and morc than 40% only in T lymphoma. Remarkably, the proportion of UCHL-1-positive cells was below the lowest value of reactive lymphadenitis just in the centroblastic–centrocytic lymphomas whose diagnosis was especially difficult on the Papanicolaou-stained smears. One NHL-T and two centroblastic–centrocytic NHL-B had been diagnosed only after incubation with UCHL-1.

However, in nearly one third of the malignant lymphomas (5/17 NHL-B, 2/7 NHL-T) the percentage of UCHL-1-positive cells overlapped the range of reactive lymphadenitis.

Benign and Malignant Lymphoid Cells in Effusions

There were only four patients with NHL in an effusion (Table 7). In the three patients with NHL-B the portion of UCHL-1-positive cells was below and the portion of L26 above that of patients with reactive lymphocytosis. In the patient with NHL-T the proportion of L26-positive cells was extremely low. As far as these few cases are representative, the results suggest that the two antibodies could be helpful in the differential diagnosis of reactive versus malignant lymphocytic effusion.

Differentiation of Reactive Mesothelia from Malignant Cells in Effusions

In the majority of the 53 cases with single atypical cells in body fluids immunocytochemistry was helpful (Table 8). Twenty-eight reactive effusions

Table 6. UCHL-1-positive cells in fine-needle aspirates of lymph nodes with malignant NHL and reactive hyperplasia[a]

	Cells (*n*)	Median (%)	Minimum (%)	Maximum (%)
Reactive hyperplasia	39	24	14	40
Malignant NHL-B	17	10	1	34
Malignant NHL-T	7	48	35	70
	63			

[a] Complemented by 51 cases from a separate study on fine-needle aspirates from fresh lymph node tissue provided for frozen section.

Table 7. UCHL-1- and L26-positive cells in effusions with malignant NHL and reactive lymphocytosis

	Cells (*n*)	UCHL-1 (%)	L26 (%)
Reactive lymphocytosis	15	25–85	4–20
Malignant NHL-B	3	1–10	90–99
Malignant NHL-T	1	85	3

Table 8. Differentiation of reactive from malignant cells in effusions with Lu-5 and HEA125 or BerEP4

	Reactive	Malignant
Confirmative	28	18
Equivocal	2	5

showed mesothelia positive for Lu-5 and negative HEA125 or BerEP4; 18 malignant effusions contained malignant cells positive to both types of epithelial antibodies. But in seven cases (five of the malignant and two of the reactive effusions) the staining results were indistinct. In seven cases the reaction was equivocal, in five because it was indistinct, in one case because of a false-positive reaction, and in one case tumor cells were scarce and could not be differentiated from mesothelial cells.

Typing of Malignant Tumors

The results of immunocytochemistry were compared to a final diagnosis based on clinical, bioptic, and/or autoptic findings in 274 cases in which

Table 9. Diagnostic yield of immunocytochemistry for tumor typing on 274 cytologic specimens

	Total (*n*)	Diagnostic (*n*)	Confirmative (*n*)	Equivocal (*n*)	Equivocal (%)
Fine-needle aspiration	179	15	105	59	33.0
Effusion	42	0	31	11	26.2
Bronchial secretion	17	2	7	8	
Brush biopsies	13	1	8	4	
Cerebrospinal fluid	11	0	6	5	
Sputum	3	0	3	0	
Other fluids	9	1	5	3	
	274	19	165	90	32.8

immunocytochemistry was performed for tumor typing (Table 2). It is not surprising that most specimens were fine-needle aspirates (Table 9), since fine-needle aspiration is frequently the first diagnostic step in tumor patients.

Tumor Typing with Immunocytochemistry in General

The results were diagnostic in only 19 (6.8%), confirmative in 164 (58.4%), and equivocal in 98 (34.8%). The diagnostic benefit was independent of the kind of specimen and of the number of antibodies used in each case. In 37 cases only one antibody was sufficient to confirm the light-microscopic diagnosis.

The 19 diagnostic results were found in fine-needle biopsies (15), bronchial secretions (two), one brush biopsy of the stomach, and one fluid in

Table 10. Contribution of immunocytochemistry to cytologic tumor diagnosis

Tumor	Total (*n*)	Diagnostic (*n*)	Confirmative (*n*)	Equivocal (*n*)
Carcinoma	107	4	67	36
Sarcoma	16	1	14	1
NHL	37	8	23	6
Carcinoid	6	1	4	1
Medullary carcinoma of thyroid gland	7	1	5	1
Carcinoma of the prostate	25	1	20	4
Hepatocellular carcinoma	13	1	4	8
Melanoma	15	0	11	4
Other tumors	48	2	17	29
	274	19	165	90

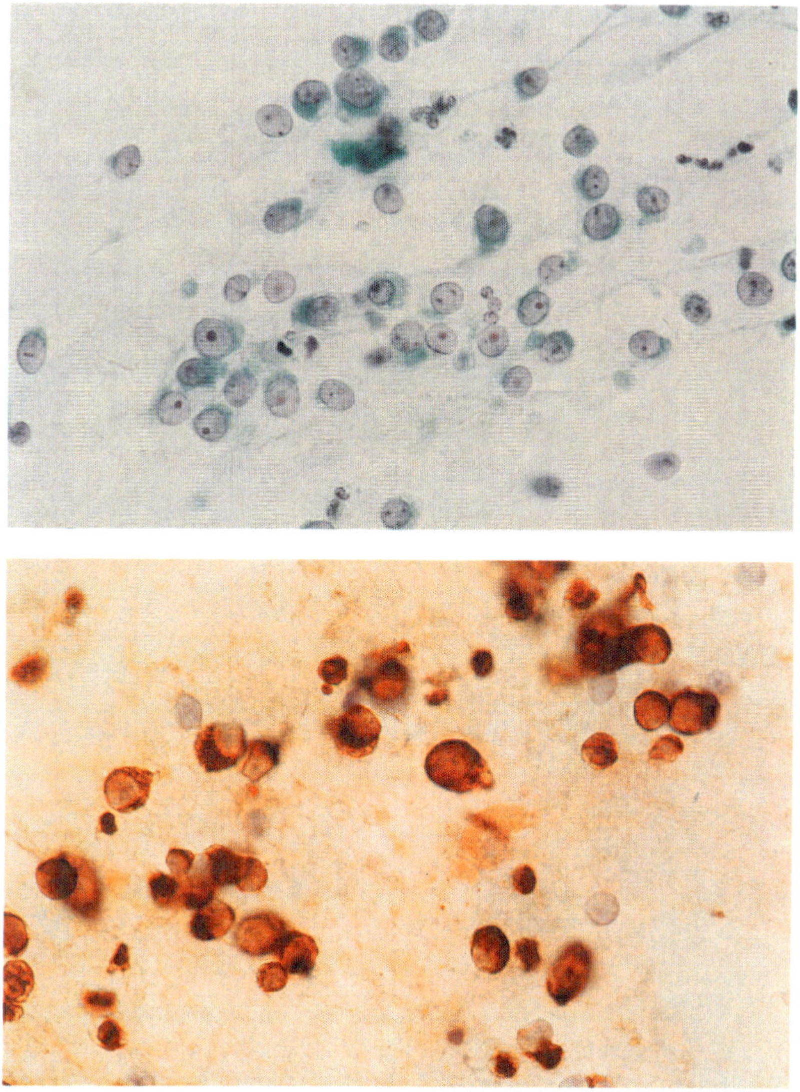

a

b

Fig. 1a–c. Brush biopsy of the stomach with neuroendocrine tumor cells (**a**, Papaninolaou's stain only) positive for vimentin (**b**) and NSE (**c**, p. 60). ×630

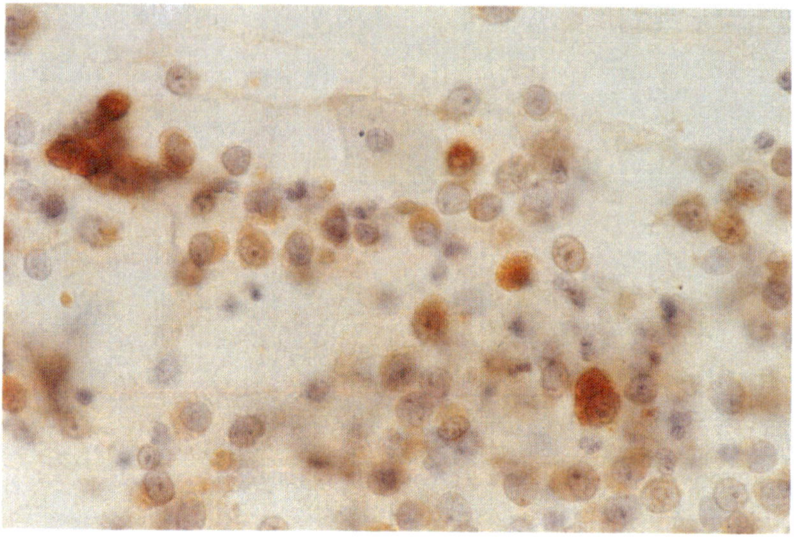

c

Fig. 1c. *Continued*

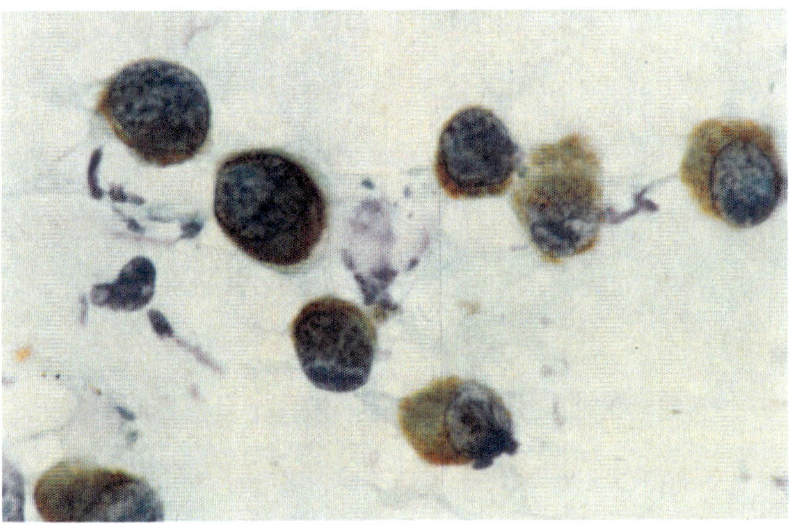

Fig. 2. Smear fixed wet with Delaunay's solution after storing in sugrose-magnesium chloride-glycerol solution at −20°C Lu-5, Papanicolaou, ×1600

Table 11. Causes of equivocal results within the whole group of the 490 cases and in the 274 cases of tumor typing

	All indications ($n = 490$)	Tumor typing ($n = 274$)
One antibody, negative staining	24	23
Several antibodies, all stainings negative	59	44
Wrong antibodies applied	19	11
Tumor cells unidentifiable	2	1
Staining results contradictory	5	3
Clearly false positive or false negative	5	0
Lack of histologic confimation	9	0
Technical failure	6	8
	129	90

which fresh tissue for frozen section was sent in. Most cases were NHL (nine) and carcinomas (five) (Table 10). In the brush biopsy of the stomach, neuroendocrine tumor cells could be demonstrated with anti-NSE (Fig. 1). In an aspiration from thyroid gland, medullary carcinoma could be diagnosed by incubation with an antibody to calcitonin (Fig. 2).

In 26.3% of all 490 cases in which immunocytochemistry had been applied the results were equivocal – most frequently because of a negative reaction and a lack of control of the antibody reactivity or a lack of internal control (Table 11). In the group of 274 cases in which immunocytochemistry had been performed as a complement to tumor typing, the results were equivocal in 32.8%. The cause of unreliable results was, in most cases, an insufficient control of negative staining, i.e., the reactivity of the antibody could not be controlled because of a lack of inner control, and the preservation of the epitopes could not be controlled by the positive reaction with a second antibody.

Differentiation of Mesothelioma from Adenocarcinoma

All mesotheliomas and adenocarcinomas reacted with the panepithelial antibody Lu-5. Three carcinomas were CEA negative, in one case the reaction was equivocal. All but one mesothelioma were CEA negative. Using only two antibodies (Lu-5, anti-CEA) 90% of all tumors were classified exactly (Table 12).

Discussion

The aim of this retrospective study was an evaluation of the diagnostic advantages of immunocytochemistry in diagnostic cytopathology. A sys-

Table 12. Differentiation of mesothelioma from carcinoma with Lu-5 and anti-CEA in 15 cases of mesothelioma and 43 cases of adenocarcinoma in effusions

	Mesothelioma	Carcinoma	All cases
Confirmative	14	39	53
Equivocal	1	4	5

tematic study on this subject with one or more clearly defined antibody panels was not possible. Many incubations were done to test new preparatory methods and new antibodies. Immunocytochemistry was not always indicated for exclusively diagnostic reasons. Therefore, it was not always easy to judge the factual value of the method, as the situation of the cytologist is quite different depending on whether the final diagnosis is known at the moment when immunocytochemistry is applied.

Though in agreement with the experience of others (Chess and Hajdu 1986; Mason and Bedrossian 1986) in up to one third of our cases immunocytochemistry was not contributory to the diagnosis, the "know how" accumulated in our laboratory during the 8-year period may improve considerably the diagnostic benefit in the future. With growing experience it became more and more clear which antibody panels are most promising in answering special diagnostic questions, and how immunocytochemistry can be successfully applied in clinical cytology.

The causes of misleading results were difficult to evaluate and often inscrutable. Lack of technical skills, however, seems not to be a major factor since we found no significant reduction of unreliable results during the last 2 years compared to the preceding period. Other causes closely related to the special quality of cytologic material such as fixation, reading problems, and problems of interpretion predominate. The limitation of cellular material makes it impossible to follow strictly the rules given in immunocytochemistry textbooks (Elias 1990; Polak and van Noorden 1983; Sternberger 1986). In particular, the main rule of controlling each incubation with positive and negative controls is difficult to observe.

Since any *fixative* may alter some epitopes, the effect of fixation on immunoreactivity has principally to be tested before an antibody can routinely be applied. Only a few studies have focused on this subject (Dinges et al. 1989; Judd and Britten 1982; Li et al. 1987). Most knowledge has to be extracted from papers dealing with immunocytochemistry with reference to special diagnostic questions.

Some examples of testing the influence of the most common fixatives to the reactivity of several antibodies are given in Table 13. For each antibody, smears from one tumor or one other tissue sample were fixed differently. We could show that the reaction of the most frequently used antibodies in

Table 13. Stability of several epitopes to six common fixatives

	Delaunay's solution (%)	Acetone (%)	Ethanol (%)	Spray (%)	Formalin (%)	Glutaraldehyde (%)
Lu-5	>90	>90	<50	>90	[2]	[2]
EMA	>50	>50	>50	>50	[2]	[3]
HEA 125	>90	>90	>50	>50	[2]	[3]
LCA	>90	>90	>90	>90	>90	>90
UCHL-1	>90	<10	>90	>90	>90	>90
L26	>90	<10	>90[3]	>90	>90	[1]
PrAP	50	>50	>50	>50	[2]	[2]
PSA	50	<50	>50	>50	[2]	[2]

[1] Strong background staining.
[2] Most cells washed off, remaining cells intensely stained.
[3] Not evaluable because of cell loss.

our material is generally good after fixation with Delaunay's solution, but may be reduced in some cases after fixation with pure solutions of ethanol or acetone.

The most common mode of fixing fine-needle aspirates is wet fixation with *spray* which preserves excellently the nuclear structures in the Papanicolaou stain and prevents washing off the cells from the slides. The immunoreactivity, however, disappears within 1–2 weeks if air-dried smears, fixed or unfixed, are stored at room temperature (Banks et al. 1983; Dalquen et al. 1986; Dinges et al. 1989; Li et al. 1987). The disappearance is accelerated if the smears have not been sprayed immediately so that cells can dry before fixation. This is the main reason for false-negative results in immunocytochemistry on fine-needle aspirates. Particularly the epitopes of Lu-5 and prostate-specific antibodies are sensitive to this kind of drying artifact. Though not absolutely inappropriate, spray fixation and transport of the dry spray-fixed smears for several days makes the results of the immunoreaction unpredictable.

Our own experience is in accordance with Judd and Britton (1982) who found *acetone*, *ethanol* or *acetone/ethanol mixtures* to give equally good fixation on cytologic material. Ethanol is used in various concentrations. Many authors prefer it to other fixatives as many epitopes are well preserved, as is the nuclear structure. At room temperature smears should not be kept in ethanol for longer than 15 days since the cells lose their immunoreactivity (To et al. 1981). Fixation with absolute or 60% aqueous acetone permits retention of immunoreactivity, especially for antibodies to cluster-defined differentiation antigens of leukocytes (Banks et al. 1983).

We prefer to use *Delaunay's solution* since with this fixative nuclear structures are brilliantly preserved in the Papanicolaou stain, and most diagnostically important antigens are not destroyed (Dalquen et al. 1986).

Moreover, we found it an ideal fixative before in situ hybridization on cytologic smears with DNA probes for Epstein-Barr virus, cytomegalovirus, and human papillomavirus. To avoid washing off the cells, the smears should be prepared on thoroughly degreased slides or briefly prefixed with spray before immersion into delaunay's solution.

Most pathologists routinely use 4% or 10% buffered *formalin* pH 7.4 to fix tissues before embedding in paraffin, since formalin-fixed material is easier to cut than cell pellets or tissues fixed with alcoholic solutions. Indeed, many epitopes are well preserved. However, formalin has also some important disadvantages: it destroys many CD antigens of leukocytes; unbuffered and highly concentrated solutions (Kung et al. 1990), solutions older than 2 weeks (Bergeron et al. 1989), and long exposure to formalin may disturb immunoreactivity of many other epitopes. Beyond that, fixation of routine cytologic smears with formalin is inappropriate because (a) many more cells are washed off from the slides when immersed into formalin than by immersion into alcohol or acetone; (b) the preservation of nuclear structure is clearly inferior to that after fixation with alcoholic solutions and acetone (Fig. 3). Therefore we think that formalin is mainly justified in cytology to fix cell pellets which are embedded in paraffin according to the cell block technique and rarely to fix smears (exception: fixation for estrogen and progesterone immuncytochemical assays).

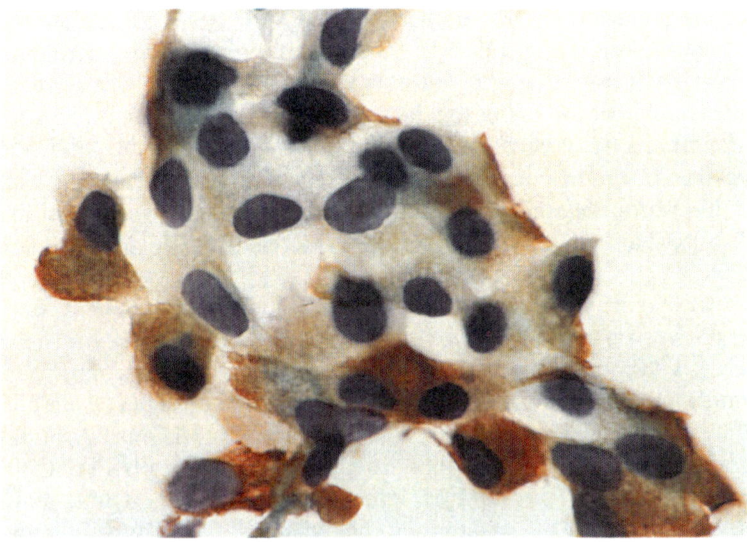

Fig. 3. Smears from a surgical specimen immediately fixed with buffered 4% formalin. Note the insufficient representation of nuclear details after formalin fixation compared to fixation with Delaunay's solution in Figs. 1, 2, 4, 5. EMA, Papanicolaou, ×400

The usefulness of *glutaraldehyde* in immunocytochemistry is disputed. After 15 min of fixation with glutaraldehyde Judd and Britten (1982) reported inadequate results for both preservation of morphology and immunoreactivity. We agree with others (Dinges et al. 1989; Leenen et al. 1985) who found that a short fixation of about 1 min does not impede immunoreactivity. Nevertheless, glutaraldehyde is inappropriate as a fixative of cell smears for the same reasons as formalin.

Many "aberrant reactions" (Chess and Hajdu 1986; Lozowski and Hajdu 1987) are due to the fact that malignant tumors do not always follow the rules of normal growth and cell differentiation. In some other cases they may be due to *reading problems*. The decision of whether a reaction is positive or negative is often difficult for a multitude of reasons, and it is not easy to find the cause of failure in a particular case.

Washing off the reaction product and the diagnostic cells during immunoperoxidase staining and restaining of the smears after incubation may occur (Chess and Hajdu 1986), but was not a major problem in our material.

After immunocytochemical incubation, *identification of the diagnostic cells* may be difficult, especially if they are very small and densely intermingled with inflammatory cells as in sputum, bronchial secretions, or serous effusions, where carcinoma cells are sometimes densely mixed up with nondiagnostic cells which are also stained.

In some cases *crowding of the diagnostic cells* gave results which could not be evaluated. Particularly the centers of dense clusters of carcinoma cells remain unstained, probably because the antibody solution is unable to penetrate the shell of the cluster (Chess and Hajdu 1986; Dalquen et al. 1986; Dinges et al. 1989). Dinges et al. observed the reverse with lymphocyte markers: in the centers of conglomerations of lymphocytes the reaction product within the small cytoplasmic rims becomes properly visible only by superimposition of the lymphocytes (Dinges et al. 1989).

Weak staining limited to the cell margin is another cause which often makes interpretation difficult. With anti-CEA, vacuolated cells may show weak staining limited to scanty extravacuolar cytoplasm. This is explained by condensation or trapping of a reaction product by surface microvilli or a component of cell membrane containing a special CEA component.

Background staining occurred in over 20% of reactions in the study of Chess and Hajdu (1986), but was not a major problem in our cases, probably because we used the ABC method. On the one hand, it may be due to high antibody concentration or nonspecific reactivity; on the other hand, staining of the debris in a smear may represent a reaction of the antibody with a specific antigen.

Degenerative changes of cells may cause false-positive staining. Sometimes it can easily be recognized by nuclear staining. In other cases it is difficult to distinguish from a true cytoplasmic reaction. But then the structure of the nucleus of the stained cells is not well preserved. Therefore, a reaction should not be judged on cells with pyknotic or otherwise damaged nuclei.

The intensity of staining may also be indeterminate in cells with *poorly preserved cytoplasm* due to inadequate fixation or prolonged drying at room temperature. In our material this was probably the main cause of inconclusive results in spray-fixed, fine-needle aspirates.

A temperature-dependent disappearance of immunoreactivity has repeatedly been shown: at room temperature or even at 4°C, fixed and unfixed air-dried preparations lose the reactivity of the epitopes of Leu 1, Leu 2a, Leu 3a, HLA-DR (Banks et al. 1983), of antibodies to cytoskeletal and other cytoplasmic proteins (Dalquen et al. 1986; Dinges et al. 1989), and of antibodies to epithelial membrane antigens (To et al. 1981) after 8–10 days. Only air-dried smears stored at −20°C retain their immunoreactivity for at least several weeks (Altmannsberger et al. 1987). One main reason for the disappearance of immunoreactivity in air-dried smears is probably dehydration of the cells. However, drying of fixed smears is probably not the only factor since immunoreactivity for Lu-5 is retained in spray-fixed smears prepared in our laboratory for at least 7 days, whereas it is frequently lost in smears fixed likewise which have come to us from other centers (see above). To preserve the immunoreactivity Dinges et al. (1989) recommend keeping fixed smears in 100% methanol at 4°C or in sugrose–magnesium chloride–glycerol solution at −20°C. Both storage methods preserve antigenicity and morphology (Fig. 2).

Lack of distinction between DAB granules and melanin granules has been claimed to raise difficulties of interpretation in peroxydase stains (Chess and Hajdu 1986). However, since immunocytochemistry is only indicated in amelanotic melanomas this is rarely a real problem.

In serous effusions where the protein content frequently exceeds that of the peripheral blood and can be well over 20 g/100 ml (To et al. 1983), the epitopes especially of immunoglobulins may be *masked* or *inactivated* by binding of fluid proteins to FC receptors and thus produce considerable staining problems (Banks et al. 1983). It is probably for this reason that we usually failed to demonstrate immunoglobulins in plasmacytes and lymphoma cells on routine preparations of effusions.

Contradictory staining results on cytologic and histologic material are most perplexing. Lozowski and Hajdu (1987) were confronted with this problem in as many as 10% of their cases. An example of this kind of equivocal result was found in the case of the neuroendocrine tumor presented in Fig. 1; at immunohistochemistry tumor cells were first negative for NSE.

Lack of controls is the second important factor which can theoretically impair the reliability of immunocytochemical studies on cytologic material. Principally, positive and negative reactions have to be controlled in cytologic as in histologic specimens. But *false-positive* reactions are sometimes difficult to control because of the small number of available smears. They can largely be excluded by stopping endogene peroxydase with H_2O_2/methanol. *Negative* reactions due to antibody inactivation can frequently be controlled by the reaction of normal cells within the smear. In other cases they can be controlled by the reaction in other specimens stained at the same time with the same antibody charge. Controls on cryostat sections are rarely needed.

Nevertheless, if diagnostic cells do not react with a certain antibody and no normal cells are present as a positive control within a smear, it has not necessarily anything to do with the differentiation of the diagnostic cells, as they could be negative in reality or negative due to a preparatory artifact. This dilemma can be partly circumvented by incubation with an alternative antibody which can be expected to react if the reaction with the first antibody is negative. We recommend, therefore, the use of at least one strong antibody like the panepithelial antibody Lu-5 or anti-vimentin. In any case, complete reliance should not be placed on one single antibody (Ghosh et al. 1983).

The experience gathered showed that the immunocytochemical examinations are performed principally for three reasons: (a) to confirm malignancy; (b) to type tumors; and (c) to assess the primary site of tumors.

Recognition of Malignant Cells

Despite the methodologic limitations recorded in the preceding section and although the number of cases in which immunocytochemistry was the only procedure leading to diagnosis was small, immunocytochemistry was helpful in the great majority of our cases. It confirmed the light-microscopic diagnosis and improved the diagnostic certainty in about two thirds of all cases. Our results show further, that immunocytochemistry can be applied successfully to any material even if most experience has been accumulated with fine-needle aspirations and effusions.

When immunocytochemistry emerged, many antibodies were thought to be tumor specific. Examples are CEA, AFP, and B72.3 (Szpak et al. 1984; Walts and Said 1983). After a while, however, it turned out that the epitope recognized by an antibody is also expressed in normal and non-malignant tissues. Therefore, the concept of diagnosing malignancy in cytologic material by a tumor-specific marker had to be given up early. Nevertheless, immunocytochemical examinations may help to detect and to confirm malignancy in some cases when malignancy cannot be diagnosed unambiguously with conventional staining methods, especially in body fluids and fine-needle aspiration biopsies from lymphomas.

Diagnostic problems may be encountered in as many as 15% of all specimens of *serous effusions* (To et al. 1981). It may be extremely difficult to differentiate between irritated macrophages or mesothelia and isolated tumor cells. Many attempts have been made to solve this problem with immunocytochemistry (Chandra and Nayar 1990; Chheda and Clarke 1986; Daste et al. 1991; Epenetos et al. 1982; Esteban et al. 1990; Ghosh et al. 1983; Gosh et al. 1986; Daste et al. 1991; Epenetos et al. 1982; Ghosh et al.

1983; Guzman et al. 1988, 1990b; Kyrkou et al. 1985; Li et al. 1989; Mason and Bedrossian 1986; Orell and Dowling 1983; Permanetter and Wiesinger 1987; Szpak et al. 1984, 1987; Tickman et al. 1990; To et al. 1981; Walts and Said 1983).

Generally the use of two (or more) antibodies has been suggested, one of which reacts with epithelial tumor cells but not with normal mesothelia and macrophages, and the other(s) with mesothelial cells only, or with both mesothelia and epithelial cells. Most authors combined an antibody against CEA with antibodies against keratins (Duggan et al. 1987; Epenetos et al. 1982; Ghosh et al. 1983), epithelial membrane antigen (EMA) (Chheda and Clarke 1986; Daste et al. 1991; Epenetos et al. 1982; Ghosh et al. 1983; Guzman et al. 1990a,b; To et al. 1981), human milk fat globules (HMFG2) (Epenetos et al. 1982; Ghosh et al. 1983) or anti-vimentin with several other antibodies (Duggan et al. 1987; Tickman et al. 1990; Walts and Said 1983). Others tried to solve the diagnostic problem with a panel of antibodies against oncofetal antigens (Orell and Dowling 1983), or against lysozyme, alpha 1-anti-chymotrypsin, tissue polypeptide antigen, keratin and carcinoembryonic antigen (Permanetter and Wiesinger 1987), or by the combination of stains for EMA, LCA (CD45), keratin and vimentin (Li et al. 1989), or by using only one antibody (Martin et al. 1986), or by the immunocytochemical demonstration of HLA-1 loss of tumor cells (Guzman et al. 1990a), or by demonstration of an FC receptor which is only expressed by macrophages but not by the cells of malignant epithelial or mesothelial tumors (Chandra and Nayar 1990).

Unfortunately, some of the strongest antibodies against epithelial epitopes (e.g., Lu-5, anti-keratin) react with mesothelial cells. Others directed against membrane-bound epithelial antigens (EMA, HEA125, HMFG2, BerEP4) are more specific and react strongly with carcinoma cells and only weakly with a few non-malignant cells. Anti-CEA, B72.3-recognizing tumor-associated globulin (TAG) and Leu-M1 binding to lacto-*N*-fucopentose, a sugar group bound to lipoproteins of cell membrane and cytoplasm, are highly specific for adenocarcinoma, but its sensitivity for epithelial tumors is only 55%–75% (Orell and Dowling 1983; Tickman et al. 1990; To et al. 1981). Therefore the value of these markers for the recognition of malignant cells is limited.

We now use the two antibodies Lu-5, which recognizes most basic and acidic cytokeratins, HEA 125 or BerEP4, and/or anti-CEA. Lu-5 is positive in the mesothelial cells but not in the macrophages of effusions and in most epithelial tumors. The HEA 125 glycoprotein, as well as EMA and HMFG2, is not usually found in mesothelia. However, interpretation of staining results is not always easy because of the weak marginal staining of these cells. Only strong staining of the whole cytoplasm should be regarded as positive (To et al. 1981).

Similar efforts were made to improve the sensitivity of *cerebrospinal fluid* cytology (Coakham et al. 1984; Li et al. 1989; Yam et al. 1987). The success depends on the cellularity of the fluid. It should contain enough cells for at least two or three cytospin preparations. Depending on the initial suspicion, we use combinations of Lu-5 with macrophage antibody Mac 387, anti-LCA, UCHL-1, L26, or an antibody against gliofibrillar antigen.

Immunocytochemical examinations are also helpful in purely *lymphocytic effusions* when the differentiation between lymphoma and reactive lymphocytosis is difficult with conventional light microscopy. The proportion of lymphocyte subtypes may also be a clue for diagnosis. In tuberculosis the proportion of CD4 is increased (Albera et al. 1991), but in neoplasms it is the same as in the peripheral blood (Guzman et al. 1990a; Spieler et al. 1986). In cases of NHL-B Spieler et al. (1986) found T cell suppression to be less than 70% and light-chain monoclonality. In the few cases of NHL we studied, the maximum of UCHL-1-positive cells was below 10% and the minimun of L26 positive cells 90% (Table 7).

The findings in *lymph node aspirates* were similar (Table 6). However, in many NHL the proportion of UCHL-1-positive cells is within the normal range. It has to be considered further that in rare cases NHL-B with a high proportion of reactive T cells may be misinterpreted as NHL-T. Therefore, in cases with a "normal" proportion of UCHL-1-positive cells, the diagnosis should not be based solely on immunocytochemistry. With reservations regarding these limitations, incubation with a T line-specific antibody may still be a valuable aid in the cytologic diagnosis of malignant NHL, especially of the centroblastic–centrocytic type in fine-needle aspirates. An alternative method is to detect monoclonality in lymph node aspirates using the polymerase chain reaction (Wan et al. 1992). Which of these methods will prevail or whether both methods will be complementary remains to be seen.

Typing of Tumors

The main task of immunocytochemistry in cytology is to help in the differentiation of tumor types. The problem is most frequently encountered in effusions and fine-needle aspirations. Many reports have appeared which document the diagnostic benefit of immunocytochemistry in individual cases. Particularly in rare tumors, the method is a valuable complement to the morphologic tools.

Domagala et al. (1989a) suggested a combination of anti-LCA and anti-vimentin for the differential diagnosis of NHL versus small cell carcinoma. We found, however, that the combination of two antibodies which recognize the same kind of cells often leads to unsatisfactory results. If one of the two possible diagnoses is only confirmed by negative staining, there is no control of the reactivity of cells and antibody.

One of the most intriguing problems is the cytologic and histologic *differentiation between mesothelioma and adenocarcinoma*. Most authors think that definite diagnosis of mesothelioma is only possible at autopsy. In this

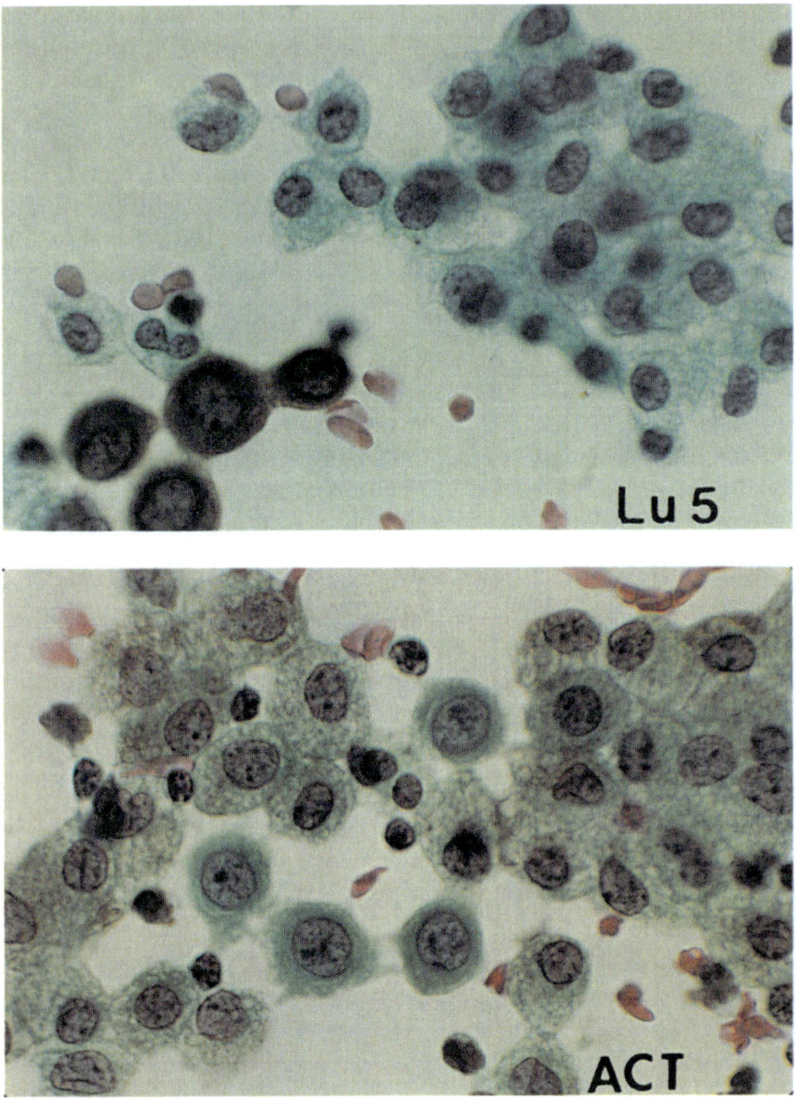

a

b

Fig. 4a,b. Normal mesothelia staining with the panepithelial antibody Lu-5, vacuolated macrophages unstained (**a**). With Papanicolaou's stain mesothelia can be distinguished from macrophages by their homogeneous cyanophilic cytoplasm (**b**). Papanicolaou, ×630 situation immunocytochemistry is a valuable complement to conventional light microscopy, though it does not solve all diagnostic problems since a mesothelioma-specific antibody is still lacking. A large number of antibodies has been tested (see Ordonez 1989).

Because of overlapping results, the diagnostic value of antibodies against intermediate filaments and associated proteins is limited. Antibodies to keratin and Lu-5 stain mesothelial surface cells (Fig. 4), nearly 100% of mesotheliomas *and* adenocarcinomas. Vimentin is positive in normal mesothelial cells and in 95% of mesotheliomas, and also in 17% of the adenocarcinomas (Duggan et al. 1987; Moch et al. 1992).

The most useful marker is CEA which is hardly ever expressed by mesotheliomas, but occurs in 75% of adenocarcinomas (Brockmann et al. 1990; Chheda and Clarke 1986; Ghosh et al. 1983; Kuhlmann et al. 1991; Moch et al. 1992; Ordonez 1989; Wang et al. 1979). Similarly the monoclonal antibody Leu-M1 recognizes 50%–60% of adenocarcinomas, but does not react with mesotheliomas (Moch et al. 1992; Ordonez 1989; Sheibani et al. 1986).

The negative staining result with anti-CEA or LeuM1 is particularly suggestive for mesothelioma if the reaction with EMA, HMFG2 (Epenetos et al. 1982; Ghosh et al. 1983; Guzman et al. 1989; Wick et al. 1990), HEA125, or BerEP4 (Moch et al. 1992; Sheibani et al. 1991) is negative. These antibodies recognizing membrane-bound epithelial epitopes are positive in 80%–100% of adenocarcinomas but rarely in mesothelioma. Therefore they are more sensitive but less specific than CEA in this differential diagnosis (Chheda and Clarke 1986). But the combination of positive reaction to vimentin and negative staining for CEA and BerEP4 or HEA125 is highly suggestive for mesothelioma as shown by our own results.

Another difficult diagnosis is *amelanotic melanoma* because its cytologic features may considerably vary from case to case. The differential diagnosis includes all kinds of dedifferentiated tumors. But the diagnostic problems can now often easily be solved with immunocytochemistry (Ordonez et al. 1988). Melanoma cells are positive for vimentin, S100 protein, and HMB45, and negative for epithelial markers with rare exceptions. The diagnosis can be made if one epithelial marker is negative and tumor cells are decorated by pAB anti-S100 or the monoclonal antibody HMB45 (Fig. 5).

Neuroendocrine tumors are now exactly diagnosed with antibodies to NSE and other neuroendocrine markers in any material. One of our cases in which immunocytochemistry first led to diagnosis was a carcinoid of the stomach (Fig. 1). Similarly a medullary carcinoma of the thyroid gland could be diagnosed with immunocytochemistry (Fig. 6).

Metastases of breast cancer and of cancers of the female genital tract may be recognized by immunocytochemical determination of estrogen and progesterone receptors. As the growth of receptor-positive cancers may be stimulated by estrogen, receptor-blocking therapy with tamoxifen is indicated. Receptor-negative tumors are not blocked by therapy. Therefore it may be important to know the receptor status when planning treatment and

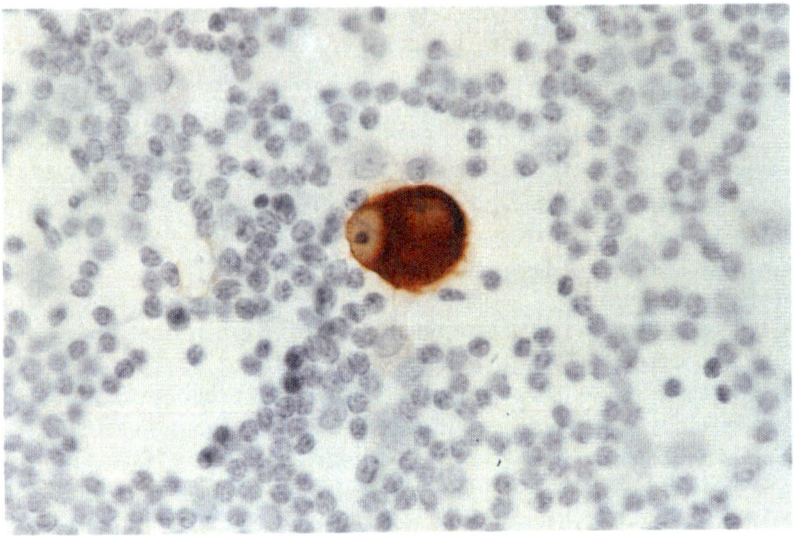

a

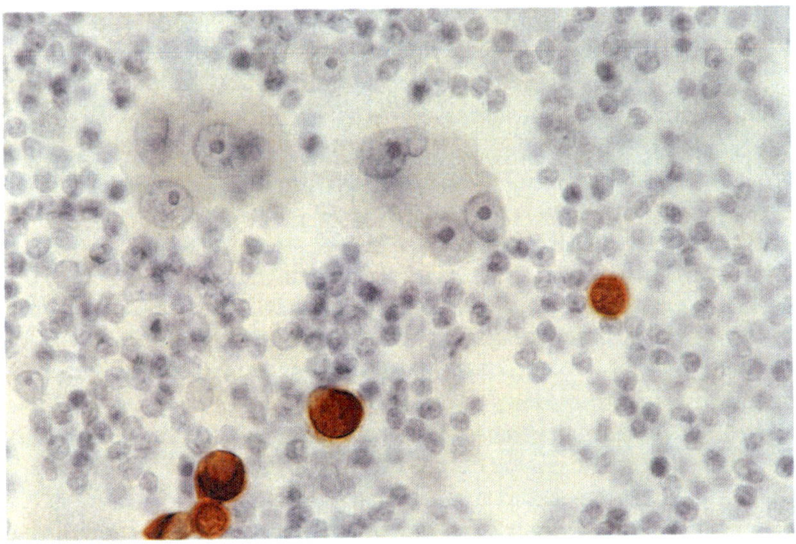

b

Fig. 5a,b. Pleural effusion with melanoma cells positive for HMB45 (**a**), negative for Lu-5 (**b**); mesothelia showing a reverse staining pattern. Counterstain Böhmer's hematoxylin, ×400

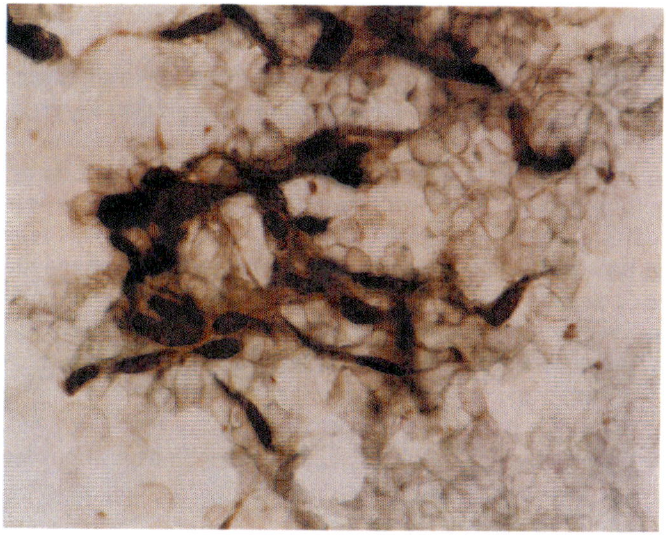

Fig. 6. Fine-needle aspiration from thyroid gland with cells of a medullary carcinoma positive for calcitonin. Spray fixation, Papanicolaou, ×400

to find the primary site of the tumor. Evaluation of the receptor status with immunocytochemical assays on fine-needle aspiration material or imprints is now well established (Deligeorgi-Politi et al. 1990; Devleeschouwer et al. 1988; Koss 1990; Vielh et al. 1992).

Hepatocellular carcinoma (HCC) may be positive for alpha-fetoprotein (AFP), hepatitis B surface antigen (HBS), CEA, vimentin, keratin, and EMA (Bedrossian et al. 1989; Carrozza et al. 1991; Domagala et al. 1989b; Thomas et al. 1989). Beyond that, a characteristic increase of endothelial cells has been shown with anti-vimentin and BMA 120 (Domagala et al. 1989b). HCC and adenocarcinoma can be differentiated with B72.3 and anti-AFP (Thomas et al. 1989). The most useful markers are probably AFP and HBS antigen. Though many HCC express AFP, the reaction was often not convincing. Others, too, emphasize that the interpretation of staining results may be difficult (Bedrossian et al. 1989).

Assessment of Primary Tumor

The hypothesis that the differentiation of a malignant tumor mirrors its histogenesis has to be modified in that, according to modern concepts of oncogenesis, all tumors independent of their origin can principally express any differentiation. Therefore, the organ-specific keratin pattern of normal cells is often not expressed in the cells of the corresponding malignant tumor. There may be at best a statistic correlation between antigenic properties of an organ and those of tumors originating from it. Very similar antigenic properties can be found in tumors from other sites. So immunocytochemical demonstration of the keratin pattern with a single antibody or special antibody cocktails are not routinely used at our laboratory.

In rare cases the origin of a malignant tumor can be inferred from special cell products. The most important example for this is the detection of metastases of *prostate carcinoma* by the immunocytochemical demonstration of prostatic acid phosphatase and prostate-specific antigen. In properly wet-fixed smears an unequivocal staining is achieved with commercially available antibodies. We could show, however, that aberrant reactions may occur in fine-needle aspirations fixed with spray after the cells have partly dried.

Another example is the proof of metstases of *ovarian carcinoma* with the antibody Ca 125 which is positive in 80%–100% of ovarian carcinomas but only in about 25% of benign serous kystomas and borderline tumors of the ovary (Neunteufel and Breitenecker 1992). Our own experience with this antibody, however, is still limited.

Other Indications

Tumor diagnosis is hitherto the main indication for immunocytochemistry in cytology. But the method now plays a considerable role in the diagnosis of non-neoplastic disorders. The most important example is the subtyping of lymphocytes in bronchoalveolar lavages (BAL). The cell composition within the BAL fluid and the helper/suppressor T lymphocyte ratio have become most relevant parameters in the diagnosis of sarcoidosis and intrinsic allergic alveolitis. A more detailed description of the special implications of BAL is beyond the scope of this article.

Summary and Conclusions

As the decision for immunocytochemistry is usually made on the basis of findings in Papanicolaou-stained smears and uncovering of the smears takes time, the immunocytochemical results are often reported with some delay. But they are of clinical interest only if reported within a short time. Therefore, immunocytochemistry on cytologic preparations must be carefully organized. The decision for immunocytochemistry must be made before the mounting medium has completely hardened to keep the time of uncovering short.

The *method* of immunocytochemistry should fulfill the following prerequisites:

1. Cell sampling and fixation should be easy to handle for the clinician who sends the specimen to the laboratory.

2. Unspecific background staining, especially in cytologic preparations rich in blood and protein, should not occur.
3. The immunostaining method should be applicable to all kinds of cytologic material, fixed and stained smears included.
4. The nuclear structure of tumor cells should not be destroyed by the immunocytochemical procedure so that tumor cells after incubation are clearly distinguishable from normal cells showing a similar reaction as the tumor cells.

There has hitherto been no such allround method fulfilling all these prerequisites since the properties of the antigenic epitopes of the cells and of the antibodies recognizing them are too heterogeneous. Therefore several methods have to be considered and a variety of technical aspects such as fixation, storage of cytologic material, properties of tinctorial stains, of antibodies and of the antigenic epitopes must be studied to find out the two or three standard methods which meet the requirements in most cases.

We recommend the ABC method for Papanicolaou-stained smears and the APAAP method for demonstration of lymphocyte markers.

The *indication* of immunocytochemistry in diagnostic cytology is restricted by the limited number of specimens. Therefore, the following rules have to be observed:

1. The conventional light-microscopic examination must have priority over the immunocytochemical examination.
2. The cytologic specimens assigned for immunocytochemical examination must have been adequately fixed and stored.
3. As the number of smears is limited, the immunocytochemical examinations must be carefully planned and restricted to the absolutely necessary incubations. If possible, an informative smear has to be spared for documentation and future training of cytologists and cytotechnicians.
4. Immunocytochemical examinations in cytology are only justified if the diagnostic problem can be clearly defined.
5. The panel of antibodies should be selected carefully so that the results may give an answer to alternative questions. At least two antibodies should be applid.

For economic reasons a laboratory occupied with routine cytology can only store a small panel of antibodies with which the most important and most frequent questions can be answered. More detailed immunocytochemical tumor typing should be reserved for large cytologic centers or laboratories integrated within an institute of pathology where many tumors are diagnosed.

The immunocytochemical results must always be interpreted together with other morphologic criteria and against a background of clinical findings. Immunocytochemistry contributes only an additional element to the morphologic findings which substantiate a particular diagnosis. Though the results are "confirmatory rather than diagnostic and seldom clarifying a primary site" (Mason and Bedrossian 1986), immunocytochemistry is now a valuable complement to cytologic methods to improve diagnostic certainty.

References

Albera C, Mabritto I, Ghio P, Scagliotti GV, Pozzi E (1991) Lymphocyte subpopulations analysis in pleural fluid and peripheral blood in patients with lymphocytic pleural effusions. Respiration 58:65–71

Altmannsberger M (1988) Intermediärfilamentproteine als Marker in der Tumor diagnostik. Veröff Pathol 127:1–105

Altmannsberger M, Dralle H, Weber K, Osborn M, Dröse M (1987) Intermediate filaments in cytological specimens of thyroid tumors. Diagn Cytopathol 3: 210–214

Banks P, Caron BL, Morgan TW (1983) Use of imprints for monoclonal antibody studies: suitability of air-dried preparations from lymphoid tissue with an immunohistochemical method. Am J Clin Pathol 79:438–442

Bedrossian CW, Davila RM, Merenda G (1989) Immunocytochemical evaluation of liver fine-needle aspirations. Arch Pathol Lab Med 113:1225–1230

Bergeron C, Shatz P, Margolese R, Major P, Ferenczy A (1989) Immunohistochemical studies with monoclonal antibodies B72.3 and MA5 on histologic and cytologic specimens from benign and malignant breast lesions. Anal Quant Cytol Histol 11:33–42

Brockmann M, Brockmann I, Fischer M, Müller K-M (1990) Reactive lesions of the pleura; immunohistochemical characterization. Pathol Res Pract 186:238–246

Brown DC, Gatter KC, Dumnill MS, Mason DY (1989) Immunocytochemical analysis of cytocentrifuged fine needle aspirates. A study based on lung tumors aspirated in vitro. Anal Quant Cytol Histol 11:140–145

Bruderman I, Cohen R, Leitner O, Ronah R, Guber A, Griffel B, Geiger B (1990) Immunocytochemical characterization of lung tumors in fine-needle aspiration. The use of cytokeratin monoclonal antibodies for the differential diagnosis of squamous cell carcinoma and adenocarcinoma. Cancer 66:1817–1827

Carrozza MJ, Calafati SA, Edmonds PR (1991) Immunocytochemical localization of polyclonal carcinoembryonic antigen in hepatocellular carcinomas. Acta Cytol 35:221–224

Chandra M, Nayar M (1990) Fc-receptors in the differentiation of mesothelial cells-from malignant cells in effusions. Acta Cytol 34:582–583

Chess Q, Hajdu SI (1986) The role of immunoperoxydase staining in diagnostic cytology. Acta Cytol 30:1–7

Chheda H, Clarke RE (1986) Immunocytochemistry: its use in the diagnosis of epithelial malignancy in serous effusions. Acta Cytol 30:586

Coakham HB, Brownell B, Harper EI, Garson JA, Allen PM, Lane EB, Kemshead JT (1984) Use of monoclonal antibody panel to identify malignant cells in cerebrospinal fluid. Lancet 1:1095–1097

Conway de Macario E, Macario AJ, Jovell RJ (1986) Slide immunoenzymatic assay for human IgE (SIA–IgE). J Immunol Methods 90:137–141

Dalquen P (1986) Diagnostischer Wert der Feinnadelpunktion (FNP) bei zervikalen Lymphknotenschwellungen. Verh Dtsch Ges Pathol 70:532–533

Dalquen P, Bittel D, Gudat F, von Overbeck J, Heitz PU (1986) Combined immunoreaction and Papanicolaou's stain on cytological smears. Pathol Res Pract 181:50–54

Daste G, Serre G, Mauduyt MA, Vincent C, Caveriviere P, Soleilhavoup JP (1991) Immunophenotyping of mesothelial cells and carcinoma cells with monoclonal

antibodies to cytokeratins, vimentin, CEA and EMA improves the cytodiagnosis of serous effusions. Cytopathology 2:19–28
Deligeorgi-Politi H, Bartziota EV, Mavrogeorgis A, Sekeris CE (1990) Determination of estrogen receptors in cytologic imprints of breast cancers using immunocytochemical methods. Acta Cytol 34:851–884
Devleeschouwer N, Faverly D, Kiss R, Legros N, de Ryckaert LY, Andry M, Lenglet G, Paridaens R, Gompel CM (1988) Comparison of biochemical and immunoenzymatic macromethods and a new immunocytochemical micromethod for assaying estrogen receptors in human breast carcinomas. Acta Cytol 32:816–824
Dinges HP, Wirnsberger G, Hofler H (1989) Immunocytochemistry in cytology. Comparative evaluation of different techniques. Anal Quant Cytol Histol 11:22–32
Domagala W, Lasota J, Chosia M, Weber K, Osborn M (1989a) Leukocyte-common antigen and vimentin are reliable adjuncts in the diagnosis of non-Hodgkin's lymphoma in fine needle aspirates. Anal Quant Cytol Histol 11:15–21
Domagala W, Lasota J, Weber K, Osborn M (1989b) Endothelial cells help in the diagnosis of primary versus metastatic carcinoma of the liver in fine needle aspirates. An immunofluorescence study with vimentin and endothelial cell-specific antibodies. Anal Quant Cytol Histol 11:8–14
Duggan MA, Masters CB, Alexander F (1987) Immunohistochemical differentiation of malignant mesothelioma, mesothelial hyperplasia and metastatic adenocarcinoma in serous effusions, utilizing staining for carcinoembryonic antigen, keratin and vimentin. Acta Cytol 31:807–814
Elias JM (1990) Immunohistopathology. A practical approach. ASCP Press, Chicago
Epenetos AA, Canti G, Taylor-Papadimitriou J, Curling M, Bodmer WF (1982) Use of two epithelium-specific monoclonal antibodies for diagnosis of malignancy in serous effusions. Lancet 2:1004–1006
Esteban JM, Yokota S, Husain S, Battifora H (1990) Immunocytochemical profile of benign and carcinomatous effusions. A practical approach to difficult diagnosis. Am J Clin Pathol 94:698–705
Ghosh AK, Spriggs AI, Taylor-Papadimitriou J, Mason DY (1983) Immunocytochemical staining of cells in pleural and peritoneal effusions with a panel of monoclonal antibodies. J Clin Pathol 36:1154–1164
Gosh AK, Spriggs AI, Charnock M (1986) Use of immunocytochemical staining to identify cells in peritoneal fluid and washings at laparoscopy and laparotomy. J Clin Pathol 39:1335–1339
Guzman J, Costabel U, Bross KJ, Wiehel U, Grunert F, Schäfer HE (1988) The value of the immunoperoxydase slide assay in the diagnosis of malignant pleural effusions in breast cancer. Acta Cytol 32:188–192
Guzman J, Bross KJ, Würtemberger G, Costabel U (1989) Immunocytology in malignant pleural mesothelioma; expression of tumor markers and distribution of lymphocyte subsets. Chest 95:590–595
Guzman J, Bross KJ, Costabel U (1990a) Malignant pleural effusion due to small cell carcinoma of the lung: an immunocytochemical cell-surface analysis of lymphocytes and tumor cells. Acta Cytol 34:497–501
Guzman J, Kraft A, Bross KJ, Wurtemberger G, Costabel U (1990b) Immunzytochemische Charakterisierung maligner Pleuraergüsse beim kleinzelligen Bronchialkarzinom. Pneumologie 44:601–602.
Hsu SM, Raine L, Fanger H (1981) The use of avidin-biotin-peroxydase complex in immunoperoxydase techniques. Am Clin Rathol 75:816–861
Judd MA, Britten KJM (1982) Tissue preparation for the demonstration of surface antigens by immunoperoxydase techniques. Histochem J 14:747–753
Kline TS, Kannan V, Kline IK (1984) Lymphadenopathy and aspiration biopsy cytology; review of 376 superficial nodes. Cancer 54:1076–1081
Koss L (1990) The future of cytology. The Wachtel lecture for 1988. Acta Cytol 34:1–9

Kuhlmann L, Berghauser KH, Schaffer R (1991) Distinction of mesothelioma from carcinoma in pleural effusions. An immunocytochemical study on routinely processed cytoblock preparations. Pathol Res Pract 187:467–471

Kung ITM, Chan S-K, Lo ESF (1990) Application of the immunoperoxydase technique to cell block preparations from fine needle aspirates. Acta Cytol 34:297–303

Kyrkou KA, Iatridis SG, Athanassiadou PP, Emmanuilidou AG, Athanassiadis PP (1985) Detection of benign or malignant origin of aszites with combined indirect imuunoperoxydase assay of carcinoembrionic antigen and lysozyme. Acta Cytol 29:57–62

Leenen PJM, Jansen AMA, Van Ewijk W (1985) Fixation parameters for immunocytochemistry. The effect of glutaraldehyde or (para)formaldehyde fixation on the preservation of mononuclear phagocyte differentiation antigens. In: Bullock GR, Petrusz P (Eds) Techniques in immunocytochemistry, Vol. III. Academic, New York, pp 1–2

Levitt S, Cheng L, DuPuis MH, Layfield LJ (1985) Fine needle aspiration diagnosis of malignant lymphoma with confirmation by immunoperoxydase staining. Acta Cytol 29:895–902

Li C-Y, Lazcano-Villareal O, Pierre RV, Yam LT (1987) Immunocytochemical identification of cells in serous effusions; technical considerations. Am J Clin Pathol 88:696–706

Li C-Y, Ziesmer SC, Wong Y-C, Yam LT (1989) Diagnostic accuracy of the immunocytochemical study of body fluids. Acta Cytol 33:667–673

Lopez-Cardozo P (1980) The significance of fine needle aspiration cytology for the diagnosis and treatment of malignant lymphomas. Martinus Nijhoff, The Hague

Lozowski W, Hajdu SI (1987) Cytology and immunocytochemistry of bronchioloalveolar carcinoma. Acta Cytol 31:717–725

Martin SE, Moshiri S, Thor A, Vilasi V, Chu EW, Schlom J (1986) Identification of adenocarcinoma in cytospin preparations of effusions using monoclonal antibody B72.3. Am J Clin Pathol 86:10–18

Mason M, Bedrossian CWM (1986) Value of immunocytochemistry in the study of malignant effusions (abstract). Acta Cytol 30:569

Moch H et al. (1992) Immunohistochemical differentiation between malignant mesothelioma and adenocarcinoma. A study on 60 cases. Virch Arch A (in press)

Nadij M (1980) The potential value of immunoperoxydase techniques in diagnostic cytology. Acta Cytol 24:442–447

Neunteufel W, Breitenecker G (1992) Distribution and localisation of Ca125, CA 19-9 and CEA in 91 ovarian tumors: an immunohistochemical study. J Tumormarker Oncol 1:99–108

Ordonez NG (1989) The immunohistochemical diagnosis of mesothelioma. Differentiation of mesothelioma and lung adenocarcinoma. Am J Surg Pathol 13:276–291

Ordonez NG, Sneige N, Hickey RC, Brooks TE (1988) Use of monoclonal antibody HMB-45 in the cytologic diagnosis of melanoma. Acta Cytol 32:684–688

Orell SR, Dowling KD (1983) Oncofetal antigens as tumor markers in the cytologic diagnosis of effusions. Acta Cytol 27:625–629

Osamura RY, Watanabe K, Yamamoto Y (1986) Immunocytochemical staining of carcinoembryonic antigen on cytologic smears. Acta Cytoll 30:88–90

Permanetter W, Wiesinger H (1987) Immunocytochemical study of lysozyme, alphal-anti-chymotrypsin, tissue polypeptide antigen, keratin and carcinoembryonic antigen in effusion sediments. Acta Cytoll 31:104–112

Polak JM, van Noorden S (1983) Immunocytochemistry. Practical applications in pathology and Biology. Wright PSG, Bristol

Ramaekers F, Haag D, Jap P, Vooijs PG (1984) Immunochemical demonstration of keratin and vimentin in cytologic aspirates. Acta Cytol 28:385–392

Sheibani K, Battifora H, JS B (1986) Antigenic phenotype of malignant mesotheliomas: an immunohistologic analysis demonstrating the value of Leu M1 antigen. Am J Pathol 123:212–219

Sheibani K, Shin SS, Kezirian J (1991) Ber-Ep4 antibody as a discriminant in the differential diagnosis of malignant mesothelioma versus adenocarcinoma. Am J Surg Pathol 15:779–784

Spieler P, Kradolfer D, Schmid U (1986) Immunocytochemical characterization of lymphocytes in benign and malignant lymphocyte-rich serous effusions. Virchows Arch (Pathol Anat) 409:211–221

Sternberger LA (1986) Immunocytochemistry, 3rd edn. Wiley, New York

Szpak CA, Johnston WW, Lottich SC, Kufe D, Thor A, Schlom J (1984) Patterns of reactivity of four novel monoclonal antibodies (B72.3, DF3, B1.1 and B6.2) with cells in human malignant and benign effusions. Acta Cytoll 28:356–367

Szpak CA, Thor A, Schlom J, Soper JT, Johnston WW (1987) Detection of adenocarcinoma in peritoneal washings by monoclonal antibody B72.3. Acta Cytol 31:649–650

Tello JW, Johnston WW (1989) A comparative study of the immunoperoxydase and immunoalkaline phosphatase staining methods using monoclonal antibodies B72.3 and antikeratin AE1,3 on cytocentrifuge specimens (abstract). Acta Cytoll 33:725

Thomas WN, Green LK, Mayberry L (1989) The application of immunoperoxydase staining to fine needle aspirates of the liver: a review of 60 cases (abstract). Acta Cytol 33:719

Tickman RJ, Cohen C, Varma, VA, Fekete PS, DeRose PB (1990) Distinction between carcinoma cells and mesothelial cells in serous effusions; usefulness of immunocytochemistry. Acta Cytoll 34:491–496

To A, Coleman DV, Dearnaley DP, Ormerod MG, Steele K, Neville AM (1981) Use of antisera to epithelial membrane antigen for the cytodiagnosis of malignancy in serous effusions. J Clin Pathol 34:1326–1332

To A, Dearnaley DP, Ormerod MG, Canti G, Coleman D (1983) Indirect immunoalkaline phosphatase staining of cytologic smears of serous effusions for tumor marker studies. Acta Cytoll 27:109–113

Vielh P, Magdelénat H, Mosseri V, Remvikos Y, di Bonito L (1992) Immunocytochemical determination of estrogen and progesterone receptors on 50 fine-needle samples of breast cancer. A prospective study including biochemical correlation and DNA flow cytometric analysis. Am J Clin Pathol 97:254–261

Van Groonen AMdVP, Herman CJ, Lindeman J (1988) Cytology immunopathology and flow cytometry in the diagnosis of pleural and peritoneal effusions. Lab Invest 58:725–732

Van Flens MJdVP, Tadema TM, Huysmans AC, Risse EK, van Meijer TGA, Meijer CJ (1990) The contribution of immunocytochemistry in diagnostic cytology. Comparison and evaluation with immunohistology. Cancer 65:2704–2711

Wadsley JJ, Watt RM (1987) The effect of pH on the aggregation of biotinylated antibodies and on the signal-to-noise observed in immunoassays utilizing biotinylated antibodies. J Immunol Methods 103:1–7

Walts AE, Said JW (1983) Specific tumor markers in diagnostic cytology: immunoperoxydase studies of carcinoembryonic antigen, lysozyme and other tissue antigens in effusions, washes and aspirates. Acta Cytol 27:408–416

Wan JH, Sykes PJ, Orell SR, Morley AA (1992) Rapid method for detecting monoclonality in B cell lymphoma in lymph node aspirates using the polymerase chain reaction. J Clin Pathol 45:420–423

Wang N-S, Huang S-N, Gold P (1979) Absence of carcinoembryonic antigen-like material in mesothelioma. An immunohistochemical differentiation from other lung cancers. Cancer 44:937–943

Wick MR, Loy T, Mills SE, Legier JF, Manivel JC (1990) Malignant epitheloid pleural mesothelioma versus peripheral pulmonary adenocarcinoma: a histichemical, ultrastructural, and immunohistologic study of 103 cases. Hum Pathol 21:759–766

Yam LT (1990) Immunocytochemistry of fine needle aspirates. A tactical approach. Acta Cytol 34:789–796

Yam LT, English MC, Janckila AJ, Ziesmer S, Li CY (1987) Immuncytochemistry of cerebrospinal fluid. Acta Cytol 31:825–833

Zajicek J (1974) Aspiration biopsy cytology, part 1: cytology of supradiaphragmatic organs. Monogr Clin Cytol 4:98–99

Nucleolar Organizer Region-Associated Proteins: A Diagnostic Tool in Cytology?

J. Rüschoff[1], K.H. Plate[2], R. Müller[3], M. Maasberg[4], and C. Thomas[3]

[1] Institut für Pathologie der Universität Regensburg, Franz-Joseph-Strauss-Allee 93053 Regensburg, Fed. Rep. of Germany
[2] Zentrum für Neuropathologie der Universität Marburg, Baldingerstrasse, 35043 Marburg, Fed. Rep. of Germany
[3] Zentrum für Allgemeine Pathologie der Universität Marburg, Baldingerstrasse, 35043 Marburg, Fed. Rep. of Germany
[4] Zentrum für Innere Medizin der Universität Marburg, Baldingerstrasse, 35043 Marburg, Fed. Rep. of Germany

Introduction

During the last 5 years silver staining of nucleolar organizer region associated proteins (AgNOR) has been introduced as a marker reaction for malignancy, and a number of reports have demonstrated a relationship between the tumours' AgNOR status and biological behaviour (for review see Crocker 1990; Rüschoff 1991).

The AgNOR technique has mostly been applied to paraffin sections. Only a few investigations have been performed in diagnostic cytology. In cell imprints of lymphoid tissue Boldy et al. (1989) could readily discriminate between high-grade and low-grade non-Hodgkin's lymphomas. In fine-needle aspirates of the breast, however, Giri et al. (1989) found a non-diagnostic overlap of mean AgNOR counts between benign and malignant lesions. In contrast, by use of digital image analysis Derenzini et al. (1988) could definitely discriminate between atypical and reactive mesothelial cells of human effusions. Similarily, Arden et al. (1989) differentiated between normal and acute lymphocytic leukemia bone marrow cells by use of image analysis. Most recently, Grotto et al. (1991) found that the distribution pattern of AgNORs is an additional diagnostic criterion that renders cell typing in smears of normal hematopoietic cells.

Since these studies were performed retrospectively or only in a small series of cytologies, the current study demonstrates our experience with AgNOR analysis in a large sample of specimens which were prospectively evaluated simultaneously with the routine diagnostic procedure. In addition, every cell was assessed both by qualitative optical criteria as well as digital image analysis in order to determine the diagnostic accuracy of different AgNOR parameters.

Recent Results in Cancer Research, Vol. 133

Materials and Methods

The AgNOR staining technique has been consecutively applied to diagnostic cytological specimens at the departments of general and neurosurgical pathology, Marburg, since the end of 1989.

Neurosurgical Specimens

Imprint preparations of 129 human brain tumours obtained by stereotaxical brain biopsy were produced by touching fresh or thawed tumour tissue on uncoated glass slides. After drying overnight imprints were fixed for 50 min at room temperature in a mixture of 70% methanol, 10% formalin and 15% glacial acetic acid and dehydrated through graded ethanols. Final tumour diagnosis and grading was done on paraffin sections of the same biopsy specimen according to the criteria of the World Health Organisation (WHO; Zülch 1979).

Effusion Cytology

Pleural (n = 55) and peritoneal (n = 13) effusions were sampled from internal medicine patients. The specimens included effusions from 45 patients with carcinoma (nine of the breast, nine of the gastrointestinal tract, seven of the lungs, eight of the bladder and the kidney, seven of the ovary and endometrium, five of unknown primary). A total of 23 patients had reactive effusions due to chronic cardiac failure (n = 12), scirrhotic liver disease (n = 5), pneumonia (n = 5) or acute pancreatitis (n = 1). Cytospin spreads were routinely performed, fixed for 10 min in 4% formalin and then rehydrated. Routine diagnostic was done by different observers in haemotoxylin and eosin and periodic acid-Schiff (PAS)-stained slides. Prior to statistical analysis all specimens were reevaluated without knowledge of the clinical and cytological diagnoses by one of us (JR).

AgNOR Staining

The silver staining was carried out in accordance with Ploton et al. (1986) using a solution of one volume of 2% gelatin in 1% aqueous formic acid and two volumes of 50% silver nitrate. The slides were incubated for 30 min at room temperature, washed in bidistilled water, dehydrated to xylene and mounted in Cargille Oil (Cargille Lab., USA).

AgNOR Evaluation

All specimens were evaluated by means of digital image analysis (Cue-2 System, Olympus Opt. Corp., Hamburg) as described previously (Rüschoff et al. 1990). A total of 30 cells were evaluated per slide. Measurements were directed to the most atypical cells exhibiting large nuclei with high AgNOR content. In the neurosurgical imprints mean AgNOR number/cell (MNORN) and mean AgNOR area per AgNOR dot (MNORA) were determined by the image analyzer. In the effusion specimens every cell was assessed optically determining MNORN and morphometrically determining the mean total silver-stained AgNOR area per nucleus (TNORA). In order to compensate staining variabilities, in every slide TNORA of ten small lymphocytes was measured ($TNORA_{Ly}$). In accordance with Derenzini (personal communication), an AgNOR index was derived by the quotient of TNORA and $TNOR_{Ly}$.

Statistical Evaluation

For statistical analysis, the Wilcoxon rank test for unpaired samples was used to compare between-group AgNOR values. The diagnostic validity of AgNOR parameters was determined in terms of sensitivity and specificity.

Results

Tumours of Central Nervous System

In general, a close relationship between tumour type and AgNOR counts was found. In neuroectodermal tumours (n = 61) MNORN increased steadily from pilocytic grade I astrocytoma (MNORN = 2.5 ± 0.5, n = 5) to grade II astrocytomas (MNORN = 3.1 ± 1.2, n = 16) to anaplastic grade III astrocytomas (MNORN = 4.5 ± 1.7, n = 10). Glioblastoma multiforme (WHO grade IV) exhibited the highest values (MNORN = 5.9 ± 2.6, n = 30). Metastases (n = 20) usually revealed diffusely distributed scattered AgNORs with MNORN of 9.6 ± 6.5 (range: 5.2–37.0). Thus, the MNORN and AgNOR distribution pattern proved to be diagnostic in differentiating between primary and secondary metastastic brain tumours (Fig. 1). Benign meningiomas of WHO grade I (n = 26) revealed low AgNOR counts (MNORN = 3.3 ± 0.3), whereas anaplastic meningiomas (n = 6) and meningeal sarcoma showed statistically significant higher values (MNORN = 5.5 ± 2.5, $p < 0.001$) (Fig. 2).

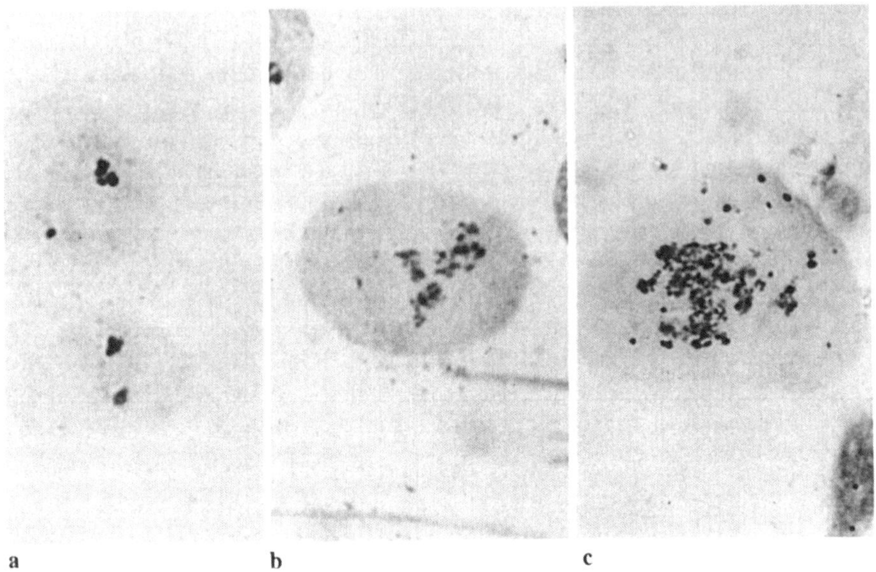

Fig. 1a–c. Few single AgNOR dots in astrocytoma WHO grade I (**a**), multiple dots mainly within nucleoli in glioblastoma WHO grade IV (**b**) and more diffusely distributed silver granules in a metastastic tumour cell (**c**). AgNOR stain, ×1000

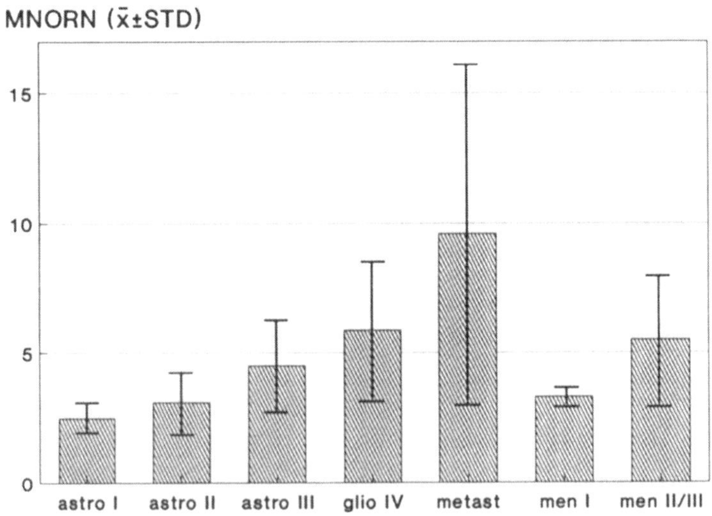

Fig. 2. Distribution of mean AgNOR number per cell in astrocytomas (*astro*, WHO grade I–III), glioblastoma multiforme (*glio*, WHO grade IV) and meningiomas (*men*, WHO grade I–III)

Effusion Cytology

Nucleolar organizer region-associated protein values were closely related to clinical and cytological data. Patients without carcinoma (n = 23) were exclusively categorized as cytologically negative both by routine diagnostic as well as AgNOR values. Based upon the reevaluation data, a total of 48 effusions were classified as negative and 20 as positive. AgNOR values were low in negative (MNORN = 11.8 ± 2.6, index = 3.0 ± 0.7) and significantly elevated in positive effusions (MNORN = 26.2 ± 15, index = 8.0 ± 4.0, $p < 0.0001$) (Figs. 3, 4). There was a good agreement between the reevaluated diagnoses and the classifications done by routine cytology and AgNOR analysis (Table 1). Thereby, the AgNOR index was most sensitive in detecting malignancy, which could clearly be demonstrated in effusions with inconclusive cytologies (Table 2). Accordingly, reevaluated diagnoses were strongly supported by the AgNOR values.

Table 1. Agreement between reevaluation diagnoses and routine cytology as well as AgNOR values

Method of evaluation	Reevaluation	
	Negative (n = 48) (%)	Positive (n = 20) (%)
Routine diagnostic	98	80
AgNOR number (>15)	90	85
AgNOR index (>4)	92	95

Values given correspond to sensitivity in positive and specificity in negative effusions.

Table 2. Diagnostic value of AgNOR parameters in inconclusive cytologies

Patient	Diagnoses			AgNOR	
no.	Clinical	Cytology	Reevaluation	Number[1] (n)	Index[2]
1	Bladder cancer	Negative	Positive	23	6.1
2	Breast cancer	Positive	Negative	7	1.5
3	Lung cancer	Negative	Positive	30	11.2
4	Unknown primary	Negative	Positive	31	9.9
5	Breast cancer	Negative	Positive	21	7.8

[1] Critical diagnostic value: >15.
[2] Critical diagnostic value: >4.

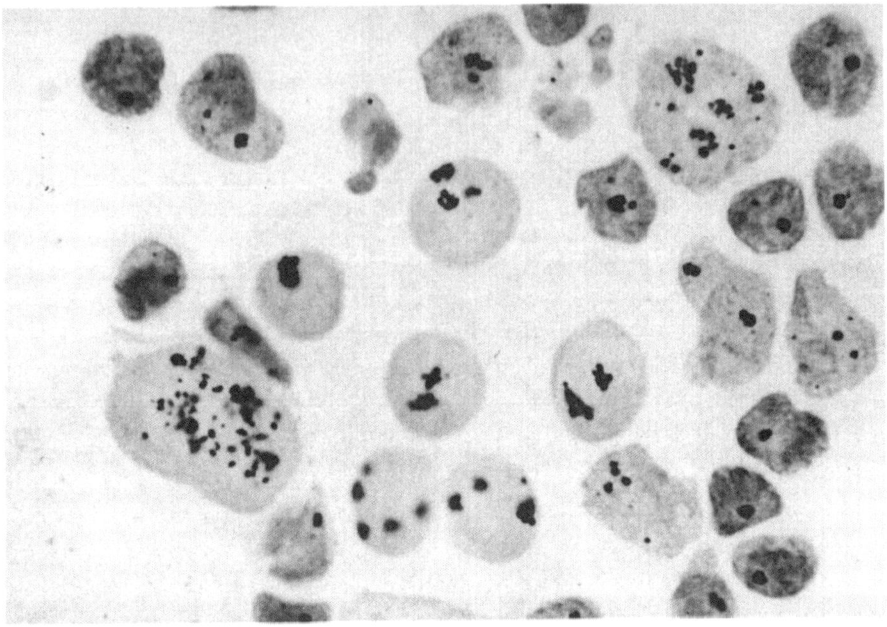

Fig. 3. AgNORs in pleural carcinosis of a breast carcinoma. Multiple lymphocytes show one or two silver-stained dots (*right*). Mesothelial cells exhibit mainly two or three AgNOR clusters (*centre*). Two carcinomatous cells are readily discriminated by enlarged nuclei with multiple AgNORs. AgNOR stain, ×1000

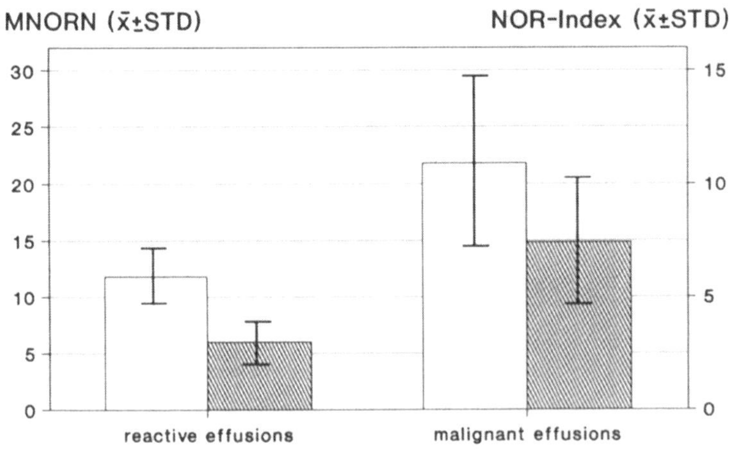

Fig. 4. Mean AgNOR number (*open columns*) and AgNOR index (*hatched columns*) in positive and negative effusions

Discussion

We describe herein that silver staining of nucleolar organizer regions is a useful tool for malignancy diagnosis in diagnostic cytology. Imprints of neurosurgically resected tumours revealed low AgNOR counts in low-grade gliomas and meningiomas. Counts in excess of 4.0 were highly suggestive of high-grade neoplasms. Metastastic lesions could readily be discriminated by most elevated AgNOR counts and scattered AgNORs. In effusion cytology AgNOR analysis contributed essential information to tumour detection in approximately 20% of cases with positive and/or inconclusive diagnoses. Sensitivity in detecting malignancy was highest if the total silver-stained area per cell was determined. Thus, the diagnostic value of AgNOR analysis in cytology was substantially improved by use of digital image analysis.

Nucleolar organizer regions are repetitive ribosomal genes (rDNA) transcribing for ribosomal RNA, which are mainly located within nucleoli of interphasic cells (Fakan and Hernandez-Verdun 1986). Activation and transcription of NORs is controlled by specific NOR-associated phosphoproteins of which polymerase I, nucleolin (C23) and nucleophosmin (B23) are the major ones (Busch et al. 1982). Silver selectively binds to these proteins; their quantity and phosphorylation status determines the amount of argyrophilic NOR staining, which is related to the proliferative activity of a given cell (Derenzini et al. 1990).

This has been confirmed by multiple studies in paraffin sections of various organ tumours (for review see Crocker 1990) and is ascertained by our results obtained in neurosurgical specimens. The MNORN increased steadily with increasing malignancy grade both in gliomas and meningiomas. Although there was a considerable overlap, values of MNORN in excess of 4 were highly suggestive for high-grade malignancies. Secondary metastatic tumours were readily recognized by significantly elevated AgNOR counts (mean value 10.0) and by a characteristically scattered AgNOR distribution pattern. Basically, these results are similar to those obtained by immunocytochemical proliferation analyses such as Ki67 immunolabelling (Plate et al. 1990). AgNOR staining has, however, some major advantages over immunostaining: (a) the technique is rapid and easily to perform; and (b) a relatively small number of cells give reliable diagnostic results.

The latter has clearly been shown in our study of human effusions. Investigation of only 30 presumably malignant nuclei revealed a high diagnostic sensitivity and specificity for both AgNOR counts and AgNOR area values. This has most recently been confirmed by Orrell et al. (1991) who found that in melanotic skin tumours a maximum of 30 evaluated nuclei gave results representative of the whole lesion.

With respect to diagnostic value, AgNOR analysis alone was as sensitive as immunocytochemistry and proved to be advantageous over DNA analysis. Immunocytochemistry, for example, yielded only a high diagnostic accuracy if a panel of antibodies was applied (Lüttges et al. 1988; Esteban et al. 1990;

Guzman et al. 1990). Using DNA flow or image cytometry, the reliabl‹ criterion of malignancy, aneuploidy, could only be demonstrated in 85% o positive effusions (Rijken et al. 1991). A combination of immunocyto chemistry and DNA cytometry has been shown to contribute either ad ditional or essential informations to tumour detection in about 17%–20% o positive or inconclusive cytologies (Croonen et al. 1988; Flens et al. 1990) An almost identical percentage has, however, been obtained in our serie: with the AgNOR technique alone. The most sensitive and specific paramete was the AgNOR index, defined as the quotient of the mean total silver stained area in presumably tumour cells (TNORA) and in small lymphocyte: ($TNORA_{Ly}$). Although the diagnostic value of TNORA has already beer shown by Derenzini et al. (1989), inevitable variations in the silver staina bility are ignored by this parameter. Therefore, the use of small restin lymphocytes as an internal staining standard has been proposed previousl (Rüschoff et al. 1990). In comparison to AgNOR counts by eye, th AgNOR index is very easily and rapidly assessed by digital image analysis In cytological specimens this paramater corresponds to the total silver stainable NOR protein quantity since whole-cell preparations are used AgNOR values in cytology are therefore higher than those obtained i histological sections (Ayres et al. 1988; Rüschoff et al. 1989).

Conclusion

Silver staining of NORs provides useful information regarding the malig nancy diagnosis in cytological specimens. The method is easily be performe and has high diagnostic validity in detecting neoplastic cells in human ef fusions. The morphometrically determined AgNOR index is a very sensitiv and specific parameter for AgNOR evaluation in diagnostic tumour cytology.

References

Arden KC, Bucana CD, Johnston DA, Pathak S (1989) Computer-assisted imag analysis of silver staining in normal and leukemic bone marrow. Int J Cance 43:395–398

Ayres JG, Crocker J, Skilbeck NQ (1988) Differentiation of malignant from norma and reactive mesothelial cells by the argyrophil technique for nucleolar organise region assciated proteins. Thorax 43:366–370

Boldy DA, Crocker J, Ayres JG (1989) Application of the AgNOR method to cel imprints of lymphoid tissue. J Pathol 157:75–79

Busch H, Lischwe MA, Michalik J, Chan PK, Busch RK (1982) Nucleolar protein of special interest: silver staining proteins B23 and C23 and antigens of huma tumor nucleoli. In: Jordan EG, Cullis CA (eds) The nucleolus. Cambridg University Press, Cambridge, pp 43–72

Crocker J (1990) Nucleolar organiser regions. In: Underwood JC (ed) Pathology of the nucleus. Springer, Berlin Heidelberg New York, pp 91–149 (Current topics in pathology, vol 82)

Croonen AM, van-der-Valk P, Herman CJ, Lindeman J (1988) Cytology, immunopathology and flow cytometry in the diagnosis of pleural and peritoneal effusions. Lab Invest 58:725–732

Derenzini M, Nardi F, Farabegoli F, Ottinetti A, Roncaroli F, Bussolati G (1989) Distribution of silver-stained interphase nucleolar organizer regions as a parameter to distinguish neoplastic from noneoplastic reactive cells in human effusions. Acta Cytol 33:491–498

Derenzini M, Pession A, Trerè D (1990) Quantity of nucleolar silverstained proteins is related to proliferating activity in cancer cells. Lab Invest 63:137–140

Esteban JM, Yokota S, Husain S, Battifora H (1990) Immunocytochemical profile of benign and carcinomatous effusions. A practical approach to difficult diagnosis. Am J Clin Pathol 94:698–705

Fakan S, Hernandez-Verdun D (1986) The nucleolus and the nucleolar organizer regions. Biol Cell 56:189–205

Flens MJ, van-der-Valk P, Tadema TM et al. (1990) The contribution of immunocytochemistry in diagnostic cytology. Comparison and evaluation with immunhistology. Cancer 65:2704–2711

Giri DD, Dundas SAC, Sanderson PR, Howat AJ (1989) Silver-binding nucleoli and nucleolar organizer regions in fine needle aspiration cytology of the breast. Acta Cytol 33:173–175

Grotto HZW, Lorand-Metze I, Metze K (1991) Nucleolar organizer regions in normal hematopoiesis: relationship to cellular proliferation and maturation. Nouv Rev Fr Hematol 33:1–4

Guzman J, Kraft A, Bross KJ, Wurtemberger G, Costabel U (1990) Immunocytochemical characterzation of malignant pleural effusions of small cell bronchial cancer. Pneumonologie 44 [Suppl 1]: 601–602

Lüttges J, Neumann K, Pflüger KH, Schmitz-Moormann (1988) Differentialzytologie von Ergußflüssigkeiten unter Anwendung von monoklonalen Antikörpern. Pathologe 9:137–142

Orrell JM, Evans AT, Grant A (1991) A critical evaluation of AgNOR counting in benign naevi and malignant melanoma. J Pathol 163:239–244

Plate KH, Rüschoff J, Behnke J, Mennel HD (1990) Proliferative potential of human brain tumours as assessed by nucleolar organizer regions and Ki67-immunoreactivity. Acta Neurochir (Wien) 104:103–109

Ploton D, Menager M, Jeannesson P, Himber G, Pigeon F, Adnet JJ (1986) Improvement in the staining and the visualization of the argyrophilic proteins of the nucleolar organizer region at the optical level. Histochem J 18:5–14

Rijken A, Dekker A, Taylor S, Hoffman P, Blank M, Kraus JR (1991) Diagnostic value of DNA analysis in effusions by flow cytometry and image analysis. A prospective study on 102 patients as compared with cytologic examination. Am J Clin Pathol 95:6–12

Rüschoff J (1991) Diagnostische und prognostische Relevanz der Analyse Nukleolus organisierender Regionen (NOR) für die pathomorphologische Tumordiagnostik. Fischer, Stuttgart (in press)

Rüschoff J, Plate KH, Bittinger A (1989) Application of the AgNOR method to cell imprints (letter). J Pathol 158:333

Rüschoff J, Plate KH, Contractor H, Kern S, Zimmermann R, Thomas C (1990) Evaluation of nucleolar organizer regions (NORs) by automatic image analysis: a contribution to standardization. J Pathol 161:113–118

Zülch KJ (1979) Histological typing of tumours of the central nervous system. In: International histological classification of tumours, vol 21. WHO, Geneva

Automation in Cytology

O.A.N. Husain

Charing Cross Hospital, London, UK

In this rapidly expanding field I will restrict my comments to those instruments dedicated to the pre-screening of cervical cancer cell samples. Where other initiatives for quantitiative cell analyses do impinge on the subject, some mention is made in passing.

The first attempt to create an automated cell scanner to read cells from the cervix was made by the Airbourne Instrument Company in the United States in the late 1950s. They constructed the Cytoanalyser, an image plane scanner using a Nipcow disc in order to analyse nuclear size and density and so identify such cancers at the pre-cancer stage (Spencer and Bostrom 1962). After spending some $4 million on it, the project failed, due largely to the limited technology of the time, but also as a result of their over-ambitious concept of hoping to find a significant difference in the overall nuclear size and density profiles when only a very small fraction of the cells would be neoplastic and of these only a much smaller fraction would be super-diploid. This problem, now known as looking for a "needle in a haystack" or a "rare event", has become a byword in this field.

For the same reason, the protagonists of the flow cell scanners failed over the years to achieve success though attempts were made to separate out the cells with higher nucleo-cytoplasmic ratios or nuclear mass, i.e., those that would be hyper-diploid (Kamentsky et al. 1965; van Dilla et al. 1967; Wheeless and Patten 1973). More recently attempts to combine flow cytometry and image analysis have been reported, but these systems are not directly applicable to cancer pre-screening targets (Ong et al. 1987).

The whole thesis of fast cell detection scanners for cervical cancer has been based on the precept that all such cancers were aneuploid. This is true for all squamous cell cancers but may not be so for all glandular tumours. As the multipotent cell of the cervical epithelium can deviate to produce glandular tumours in round 5%–10% of the whole this single parameter scan may well cause limitations for such an approach.

Nevertheless, many initiatives have resulted in prototype scanners over the past 20 years based on the concept of an interactive scan at least for the

Recent Results in Cancer Research, Vol. 133

first-generation machine where the cytologist makes a decision on that rare event of a possibly neoplastic cell amidst a mass of normal cells. Apart from an attempt at a rotating mirror scanner on cells crudely tracked on the Mylar tape and stained by Papanicolaou's technique (Spriggs et al. 1968), the general pursuit has been towards television, cathode ray and diode array scanners.

We ourselves, in 1965, were first in converting a vidicon scanner, the Quantimet B Image Analysing Computer, from an incidental light microscopic scanner used for industrial purposes to a transmission light microscope able to read the size and density (integral optical density) of Papanicolaou-stained nuclei in cervical scrape smears using their size and density potentiometers and an automatic stage.

We demonstrated that this was a feasible approach with enough discrete nuclei to achieve an adequate register but also showed that a substantial number of abnormal nuclei were buried within clusters and overlaps and therefore inaccessible to the scanner (Husain and Henderson 1970). The British Ministry Of Health then funded a substantial research programme for us to look into cell preparation methods as well as to build another scanner.

Cell Preparation Techiniques

To achieve a maximum register of cells for quantitative assessment, we set to work throughout the early 1970s to investigate means of liberating the "buried" cells and to present them for easy and rapid scan analysis. We explored:

1. Improved sampling instruments and this resulted ultimately in the Aylesbury spatula (Wolfendale et al. 1987) to achieve a better sample from the junctional zone of the cervix.
2. The choice of an optimal sampling and preservation solution for cell transport – here a 10% alcoholic normal saline appeared optimal after trying out enzymic and chemical ingredients.
3. Disaggregation systems by chemical, enzymic and physical methods, the latter including ultrasound and sheer stress systems using syringing peristaltic pumps and vortex whistles (Husain et al. 1974) – these were all designed to separate out the secondary cell aggregates but not to disrupt cells that had grown together; to do that would have meant disrupting the desmosome links which would have resulted in disintegration of the cells themselves and the creation of cell debris to further confuse the scanners.
4. The next requirement was for cell marshalling onto some transparent surface and, though plastic film strips and Mylar tapes were explored, the clarity and refractive index of glass ensured that microscope slides would be the choice.

5. Methods of adhering disaggregated cells from watery substrates resulted in our developing a poly-L-lysin method by coating a slide with a 1% aqueous solution and letting it dry when it presents free positive charges on its surface. Any cell suspension laid on such a surface would allow the negatively charged cells in solution to form a negative–positive bond and, on washing off the excess fluid and cells after some 3–5-min sedimentation, a non-overlapping monolayer would result in fast and easy scanning (Husain et al. 1978; Watts and Husain 1984a).
6. The next need was for a stable stoichiometric stain as the spectral peak for haematoxylin was lost in a few weeks. Having reviewed some 30 different nuclear and cytoplasmic stains, we settled on modifying one and so we developed a 2–5-min gallocyanin stain from the Einerson 18-h schedule (Husain and Watts 1984). Such a stoichiometric nuclear stain could replace haematoxylin in the Papanicolaou schedule as well as in histological stains and would be appropriate for the increasing demand for quantitative microscopy today. Such stains are being assessed and standardised (Wittekind 1985; Schulte and Wittekind 1990).
7. Finally there was need to build a non-contaminating single-slide staining machine especially for use of monolayers where cross-contamination between smears cannot be easily identified. However, although I have presented two designs for consideration over the past 23 years, no company has been prepared to construct a safe one. Our report on cross-contamination by bulk staining and one inverted slide stainer has demonstrated that anything between 15% and 100% cross-contamination of cells can occur (Husain et al. 1978b).

A brief mention of the highly optimised microscope environment (HOME) project is appropriately considered here. This is a European collaborative venture which has resulted in a quantitative microscope for use by all future quantitatively minded pathologists and cytologists on which accurate special measurements as well as densitometric values can be achieved. The HOME microscope presents all the necessary investigational and experimental measurements with histogram data relating to cell and tissue clearly depicted in the microscope's field of vision thus making it unnecessary to refer to paperwork (von Hagen et al. 1991).

Returning to our subject of the fast automatic pre-screening scanners, the British Ministry of Health in 1969–1970 funded the Imanco (Quantimet) Company to construct a purpose-built scanner, the Cytoscreen, to my own specification. This was a plumbicon scanner because of its virtue of fast erasure of images which allowed the instrument to scan some 11 frames (or 300–500 cells) per second. Used on a raster scan with its ability to size, shape and count the over-threshold nuclei, it became one of the few test beds for the collaborative meetings of multidisciplinary groups in New Hampshire and California through the 1970s under the generous patronage of the American Engineering Foundation.

Regrettably, the government's funding of the Cytoscreen project ceased, and the Medical Research Council's Cytoscan (later called Cervifip), an offshoot from a cytogenetic scanner development for finding metaphase spreads and counting chromosomes, was concentrated on by British workers at Edinburgh and Charing Cross Hospitals.

The Cytoscan/Cervifip was a charge-coupled linear diode array using specially prepared monolayer smears, initially by the poly-L-lysine technique and later the nucleopore impress smear developed in Nijmegen (Zahniser et al. 1982). Results of years of development of the Cytoscan are published (Husain et al. 1992) but now with the addition of a high-resolution package, an automated slide feeder and improved software the Edinburgh scanner is reaching competetive levels with other projects.

A brief account of other projects follows.

The *BioPEPR* project, developed in the Institute of Pathology and Physics Laboratory, University of Nijmegen, was based on high-resolution slide photography and the use of nucleopore transfer of cells in solution to a glass slide by alcoholic compress (Zahniser et al. 1982).

The CYBEST, now at a series 4 stage, is a high-resolution automated cytological screening system utilising image analysis on cells prepared on a monolayer smearing machine, the Cybest-CDMS developed by the Toshiba research and development centre, Kawasaki, Japan (Tanaka et al. 1987).

The DIASCANNER was developed by IMTEC Image Technology in Uppsala between 1976 and 1987. As with the Cytoscan/Cervifip, it was targetted on both pre-screening and metaphase findings (Nordin et al. 1989).

The *CYTYC* Corporation in Marlborough, Mass, has produced a much more sophisticated instrument wherein, by scanning a mildly disaggregated cell sample of a thin-film preparation, the apparatus selects some 64 segment frames on the high-resolution monitor screen together with further rows of "normal cells", "cell clusters" and inflammatory cell patterns with the additional facility of magnification of suspect cell images and the creation of histograms of cell parameters superimposed on the monitor face to assess and support a suspicious signal. This instrument is currently undergoing trials in the United States (Hutchinson et al. 1991).

The FAZYTAN is a system using optimised television microscopy with a host mini-computer and peripheral array processors and digital image stores developed by the University of Stuttgart (Schwarzmann et al. 1989).

The LEYTAS is a fast television image parallel processing computer developed in the Departments of Cytogenetics and Histochemistry and Cytochemistry, University of Leyden (Ploem et al. 1982).

The TULIPS is a system built around a high-speed data-bus (40 Mb/s) connected to a PDP-11/55 host computer. An image memory, an LSI-11 microcomputer and a video interface are also connected to the high-speed bus. The system was developed at the Departments of Computer Science and Physics and of Clinical Cytology, Uppsala University Hospital, Sweden (Stenkvist and Strande 1989).

ACCESS is a fully automated cervical cell pre-screening system produced by the Xillix Company and the British Columbia Cancer Agency. They process each smear by employing the Feulgen $TININESS_2$ stain to scan for neoplastic cells using a CCD condenser and record the co-ordinates of each suspect cell which can be confirmed by subsequent Papanicolaou staining. They hope to exclude a substantial proportion of smears as negative thus saving manual microscopy (Palcic et al., in press).

The ROCHE/KONTRON is a joint production development by the University of Hannover and Kontron of Munich. Real colour images are evaluated and automatically stored for retrieval to be viewed by eye. The use of the Roche Autocyte cytorich cytology preparation is to provide a more efficient scan of cells.

The PAPNET is produced by Neuromedical Systems Inc. of Suffern New York. It is a sophisticated machine utilising algorythmic and neural network analysis on scans of standard Papanicolaou-stained cervical spatula scrape smears. This permits trials on archival material which can be examined and reported on. This system reduces the viewing to 64 or 128 of the most abnormal cells or patterns on one or two monitors which the cytologist can scan rapidly. Most of the smears can be excluded from full smear scan in this way and this should relieve the cytotechnologists of the tedium of massive routine screening and/or also serve as a method of quality control. Both aspects are currently being evaluated by a number of independent authorities.

The AUTOPAP 300 by Neopath also reads Papanicolaou-stained cervical scrape smears and classifies each smear as normal or requiring human review. This can be carried out by an ordinary microscope or by their Autoreview microscope.

Further projects, mostly in the United States, are developing rapidly and companies such as the AMCELL 2000 T Laboratory are developing fast and should be watched closely.

It is thus possible to see the wide extent of advance that has occurred in the field and, though many initiatives have failed to come to fruition, they have all contributed to the continued development of knowledge, expertise and both hardware and soft-ware dedicated to this field. It is expected that trials of one or two are imminent and that a commercial product will be in some sort of parallel use within 5–10 years. This will come just in time as the demand for cervical cancer screening is not being met all over the world and, as the machines and systems improve, other cancers such as lung and urinary tract are obvious front runners. Even if these machines get adopted, it is unlikely that a significant reduction of man power will result. The technicians will only become more skilled and high powered in handling such sophisticated systems.

References

Husain OAN, Henderson MJ (1970) Observations on the use of the Quantimet image analysing computer in automatic scanning for malignant cells. Proceedings of the 2nd Tenovus Symposium, Cardiff 24–25 Oct 1968, pp 214–227

Husain OAN, Watts KC (1984) Rapid demonstration of nucleic acids using "oxidised" gallocyanin and chromic potassium sulphate: methods and applications. J Clin Pathol 37:99–101

Husain OAN, Allen RWB, Hawkins EJ, Taylor JE (1974) The Quantimet Cytoscreen and the interactive approach to cancer screening. J Histochem Cytochem 22:678–684

Husain OAN, Page-Roberts BA, Millet JA (1978a) A sample preparation for automated cervical cancer screening. Acta Cytol 22:15–22

Husain OAN, Grainger JM, Sims J (1978b) Cross contamination of cytological smears by bulk automated methods. J Clin Pathol 31:63–68

Husain OAN, Watts KC, Lorriman F, Butler B, Tucker J, Carothers A, Eason P, Farrow S, Rutovitz D, Stark M (1993) Semi-automated cervical smear prescreening systems: an evaluation of the Cytoscan-110. Analyt Cell Pathol 5:49–58

Hutchinson ML, Cassin CM, Ball HG (1991) The efficacy of an automated preparation device for cervical cytology. Am J Clin Pathol 96:300–305

Kamenski LA, Melamed MR (1965) Spectro-photometer: New instrument for ultra rapid cell analysis. Science 150:630–604

Nordin B, Bengtsson E, Eriksson O, Jarkrans T, Stankvist B. Some experiences from our cervical prescreening project. Proceedings of the 1st conference of the European Society for Analytical Cellular Pathology, 12–17 Nov 1989, pp 145–146

Ong S-H, Horne D, Yeung C, Nickolls P, Cole T (1987) Development of an imaging flow cytometer. Analyt Quant Cytol Histol 9:375–382

Palcic et al. Analyt Quant Cytol Histol (in press)

Ploem JS, Goyarts-Veldstra L, van Driel-Kulker AMJ (1982) Potentialities of Leytas in analytical and quantitative cytology and its specific applications in cervical cancer screening. Analyt Quant Cytol 4:155

Schewartzmann PW, Sprenger E, Dorrer R, Fischer JE, Greiner W, Schlipf W (1989) Prescreening experiments with Fazytan-IPS. Proceedings of the 1st conference of the European Society for Analytical Cellular Pathology, 12–17 Nov 1989, pp 139–140

Schulte E, Wittekind D (1990) Standardisation of the Papanicolaou stain I. A comparison of the five nuclear stains. Analyt Quant Cytol Histol 12:149–156

Spencer CC, Bostrom RC (1962) Performance of the Cytoanalyzer in recent trials. J Natl Cancer Inst 29:267–276

Spriggs AI, Diamond RA, Meyer EW (1968) A new principle for automated screening of cervical smears. Lancet i:359–360

Stenkvist B, Strande G (1989) Analysis of machine-selected cells with an image analysis system in normal and abnormal cervical specimens. Analyt Cell Pathol 2:1–13

Tanaka N, Ueno T, Ikeda H, Ishikawa A, Yumauchi K, Okamoto Y, Hosoi S (1987) Cybest model 4: automated cytologic screening for uterine cancer utilizing image analysis processing. Analyt Quant Cytol Histol 9:449–454

van Dilla MA, Fulwyer MJ, Boone IV (1967) Soc Exp Biol Med 125:367–372

von Hagen V, Morens A, Krief B (1991) HOME: Highly Optimised Microscope Environment (a new work station for microscope analysis in pathology). Analyt Cell Pathol 3:249–256

Watts KC, Husain OAN (1984) Optimal use of the cationic polyelectrolyte poly L lysin in the preparation of cell monolayers for diagnostic cytopathology. J Clin Pathol 37:829–835

Wheeless LL, Patten SF (1973) Slit-scan cytofluometry. Acta Cytol 17:333–339
Wittekind D (1985) Standardisation of dyes and stains for automated cell pattern recognition. Analyt Quant Cytol 7:6–31
Wolfendale MR, Howe-Guest R, Usherwood MM, Draper GJ (1987) Controlled trial of a new cervical spatula. Br Med J 294:33–35
Zahniser DJ, Oud PS, Raaijmakers MCT, Vooys GP, van de Walle RT (1982) The BioPEPR system for cervical cytology automation. Analyt Quant Cytol 4:164

II. State of the Art of Old Methods

Sampling and Processing in Gynecological Cytology

M. Hilgarth

Universitäts-Frauenklinik, Hugstetterstr. 55, 79106 Freiburg, Fed. Rep. of Germany

Cytological diagnosis is a procedure of several, often numerous steps. Obviously, the quality of such a multistep process depends ultimately on the weakest link in the chain. In this context, the sampling and preparation techniques are the most frequent problems. The rates of false-negative cases in routine cytology of the uterine cervix may be as high as 15%–30%, estimated to be even higher under pessimistic aspects. From our own studies, which have been confirmed worldwide, we realise that about 60%–70% of these false-negative findings are due to methodological errors, i.e., absence of atypical cells in the smears. This may depend on the size of the tumor, on its occult localisation in the cervical canal or behind a vaginal stenosis. Another problem is how far spontaneous exfoliation of a tumour may be impeded by keratinisation. Moreover, the diagnostic results depend on the sampling apparatus and on the aggressivity of smear taking.

The prerequisites for differential cytology are listed in Table 1, placing particular emphasis on contact smears from the portio and cervical canal. They must accurately represent the entire epithelial pattern on the slides prepared. The number of cells in the smear must be sufficient for evaluation. The smears must not be mixed or blended, but should be put carefully and gently on the slide with the appropriate instrument.

Above all, the material taken must never be homogenised or applied in a thick coat. Previous and current tests in view of automatisation have illustrated the problems of clustered cells for any evaluation, even with the naked eye. "Bad smears" in this respect will be too dense, too thick, with too much blood, and in partial or total autolysis. Figures 1 and 2 show examples of bad smears which are insufficient for diagnostic evaluation.

Sampling of cytological smears was initiated by Papanicolaou with aspirates from the posterior vaginal vault using a pipette. This means that he collected exfoliated cell material accumulating from the vaginal pool. The advantage of this procedure lies in the inclusion of cells from the upper genital tract. It is currently used in the United States where separate smears are taken from the endocervix and ectocervix, with an additional smear

Recent Results in Cancer Research, Vol. 133

Table 1. Prerequisites for differential cytology

1. Only those contact smears from the portio or cervical canal which represent the entire extent of epithelial change should be used for evaluation.
2. The smear must be readily evaluable as a result of adequate fixation and staining, and absence of inflammatory side phenomena.
3. The cytologist must be familiar with the interpretation of cell types.
4. Corresponding histological material must be available for comparative examination.
5. The number of cells on the smear must be sufficient.

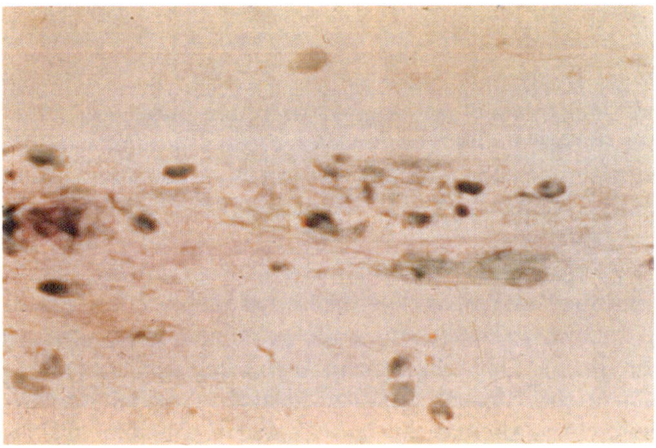

Fig. 1. Autolytic cell material unfit for differential evaluation

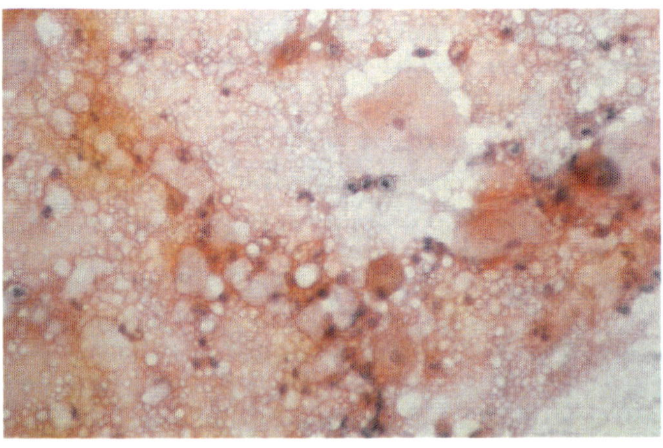

Fig. 2. Autolytic cell material poorly preserved and unfit for evaluation

from the vaginal pool, offering the possibility of detecting endometrial carcinoma by comparably simple means. Ayre (1947) and Graham (1947) abandoned aspirates from the vaginal pool in favour of direct smear sampling. The border between squamous and columnar epithelia, the site of carcinogenesis, was approached directly using the Ayre spatula or other instruments.

In Germany, the platinum hook was initially used to take endocervical smears, and a cotton wool-tipped rod and the wooden Ayre spatula was used to take portio smears. The precoiled cotton wool tip, commercially available, represented considerable progress since it could be used for endo- and ectocervical smears. Its drawback, however, is that it cannot transfer more than a minimum of cells onto the slide since the tip is covered with a special coating necessary for industrial precoiling and preparation. Unfortunately, up to 80% of the exfoliated cells will stick to the fibers and escape transfer to the slide itself. Repeated attempts have been made to develop better instruments to solve the problem: for example, there was the pipe cleaner, much thinner than a cotton wool-tipped rod, which would favour sampling even from a narrow os uteri as in older women. Moreover, the transfer and "rolling" of cells on the slide was improved.

Another instrument was developed to collect material from endo- as well as ectocervix by one swab with plastic fibers. The idea of this tool, called Cervex, appears plausible, but after testing it in some 300 cases, we failed to achieve adequate results. The instrument is poorly adaptable to the individual anatomical conditions, and bleeding is frequent.

Some gynaecologists are using thin platinum rods with cotton wool tips to enter a stenosed cervix. A large spectrum of sampling instruments is now available, but none of these tools meets all our requirements.

Eventually Szalay (Hilgarth and Szalay 1986) developed a specific spatula to collect material from endo- and ectocervix in a single procedure. The tip of this spatula, introduced into the cervical canal, is readily adaptable to the individual condition (Fig. 3). Smear taking begins by placing the "shoulder" of the curved spatula firmly onto the portio. Then the spatula is rotated through 360°, with repetition if necessary, withdrawn, and the collected material is smeared carefully on the slide. Now the upper part of the slide is coated with the endocervical, the lower with the ectocervical material. The spatula is made of plastic material with a knobbed surface to stimulate exfoliation in a more aggressive procedure, yielding not only spontaneously released superficial cells, but also those from deeper layers.

The cervical brush proper, the so-called cytobrush (Fig. 4) permits entering the cervical canal without pain or problems. The surface of the brush favours aggressive collecting from deeper cervical crypts, with excellent transfer to the slide because fewer cells are retained in the brush than in cotton wool tips. Current tests are evaluating a smaller version, the "mini-cytobrush", to be used in cases of genuine cervical stenosis in older women. Another modification foresees that the metal tip, which is often

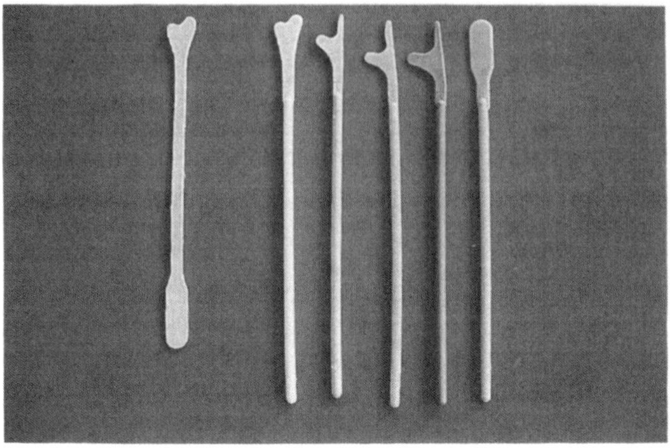

Fig. 3. *Left*, Ayre's wooden spatula. *Right*, five models of the Szalay spatula

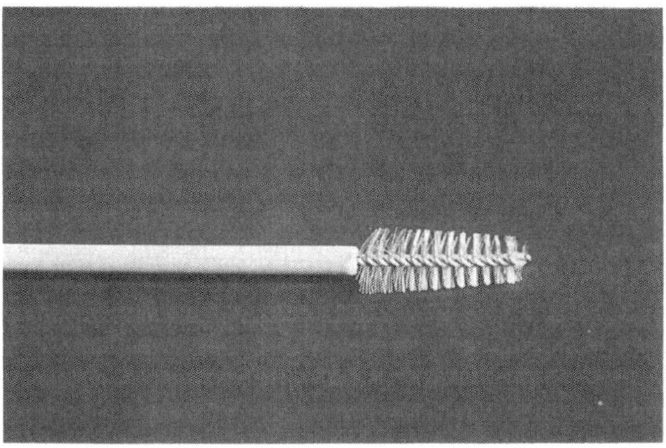

Fig. 4. Cytobrush

liable to cause bleeding, be protected by an oval plastic button. The latter would also serve to find the way into a stenosed canal and so avoid a false entry.

It is still an open question whether and how "a smear from the cytobrush would suffice to document the free passage of the cervical canal without histologic findings, i.e. without endocervical curettage?". The problem is of the utmost significance, for instance in a case of dysplasia prior to the consideration of conservative therapy such as cryotherapy or laser therapy. Above all, it is a legal problem, repeatedly arising in medical and legal expertise when no endocervical curettage has been performed, but only an endocervical smear has yielded negative cytological findings.

Let me briefly review the conditions for endometrial sampling. Some 20 instruments are commercially available to date, but none are fully satisfactory. The main problems are:

- Entrance via the interior os uteri involves considerable pain for the patient.
- How much representative accuracy can be attributed to the collected material with regard to the large uterine cavity with its poorly accessible tubal corners?
- How can the collected material be handled and prepared: by smears, by thrombin cell blocking, by paraffin blocks?
- Who is sufficiently trained for correct evaluation of this complicated endometrium cytology from direct sampling?

These are the main questions that are currently under controversial discussion.

In my own opinion these questions have lost some of their relevance since the introduction of diagnostic ultrasound. In patients at risk it is now possible to locate exactly any endocavital changes and to recommend definitive clarification. Mucosal thickness can be measured with great accuracy, and the result will be decisive regarding the necessity for abrasion. In doubtful cases, ultrasound controls are indicated after tamoxifen therapy. Of course, any reliable cytological diagnosis, in both cervical and endometrial problems, requires adequate documentation of the relevant anamnestic data from the sender. Staining is performed according to Papanicolaou by hand or using an automatic processor; stains of high quality and purity are indispensable, and correct staining times are most important. In our laboratory, pre-embedding dehydration is still performed with xylol, for which we have never found an adequate substitute. In fact, the substance is not as dangerous as it is often supposed to be provided the necessary precautions are respected.

To date, errors in preparation and staining have become rare due to improved technical skill and equipment.

Conclusion

According to our own experience, the method of choice for cervical cytology is the separate collection of smears from the ecto- and endocervix. The choice of instruments depends on the facility to correctly represent the epithelial border, preferably after colposcopy. The borderline between squamous and columnar epithelium has to be exactly located: if it is found in the ectocervical part, the wooden spatula or a similar tool is indicated. In endocervical sites, the cytobrush has been an adequate instrument in our laboratory.

References

Ayre JE (1947) Selective cytology smear for diagnosis of cancer. Am J Obstet Gynecol 53:609–617

Graham RM (1947) The effect of radiation on vaginal cells in cervical carcinoma. Surg Gynecol Obstet 84:153–166

Hilgarth M, Szalay L (1986) Farbatlas der gynäkologischen Zytodiagnostik. Marseille Verlag, Munich

The Munich Nomenclature

H.J. Soost

Institut für klinische Zytologie, Technische Universität München,
Prinzregentenplatz 14, 81675 Munich, Germany

When the new developments in gynecologic cytology became known in German-speaking countries after World War Two, Papanicolaou and Traut (1943) had established the classification in five groups. Older colleagues who witnessed those years may remember negative evidence, group III – unclear, groups IV and V – positive findings. In particular group IV reflected the presence of some isolated abnormal cells, group V that of numerous abnormal cells in the smears.

During the introduction of cytology, this system by Papanicolaou had been most useful. However, it was derived from the prevailing concept that carcinoma could be diagnosed only from coherent tissue, and not from isolated cells. Moreover, Papanicolaou, not being a pathologist but an anatomist, had to be extremely careful in the interpretation of his findings in order to avoid more controversies than he had had to endure over the years. We have witnessed similar conflicts in Europe, e.g., in 1946 when IGEL, reporting his first investigations in gynecological cytology to the first Congress of Gynecologists in post-war Germany, met with severe attacks: one of the leading German pathologists spoke of a "bad Care parcel received as a gift from the USA."

Today, everybody knows the close correlation between cytologic and histologic findings, but this knowledge has taken many years to be accepted.

Since the mid-1960s, many ideas have been proposed to give a detailed morphologic identification of Papanicolaou groups, based on histologic data, but these proposals varied widely from one laboratory to the other.

In 1973, a WHO recommendation proposed abandoning group classification in favor of the exact description of findings in adequate terminology as used in histologic records (Riotton et al. 1973). Although the concept was supported by the International Academy of Cytology, it has gained only restricted acceptance worldwide.

The poor response has been due, in my opinion, not only to a certain resistance to verbal description as such, but also to the inadequacy of the histologic WHO classification – light, moderate, severe dysplasia, micro-

Recent Results in Cancer Research, Vol. 133

Table 1. Recording in gynecological cytodiagnostics (Munich Scheme of 1975; Soost and Baur 1990)

Group	Cytologic findings	Recommendations
I	Normal cell pattern	
II	Inflammatory, regenerative, metaplastic, or degenerative changes, hyper- or parakeratotic cells	Possibly repeated smear
III	Severely inflammatory or degenerative changes and/or poorly preserved cell material; dysplasia, carcinoma in situ, or invasive carcinoma not excluded; abnormal glandular or stromal cells of postmenopausal endometrium	Short-term cytologic monitoring necessary after hormone or antibacterial treatment, evtl. histologic check up
IIID	Cells of mild or moderate dysplasia	Cytologic check up after 3 months
IVA	Cells of severe dysplasia or carcinoma in situ	Histology
IVB	Cells of severe dysplasia or carcinoma in situ, invasive carcinoma not safely excluded	Histology
V	Cells of invasive cervical carcinoma or of other malignant tumors	Histology
O	Technically inadequate (e.g., sample of insufficient quantity, poor fixation)	

carcinoma, clear cell mesonephroid carcinoma etc. – for an appropriate correlation with cytology. Even experienced senior cytologists might be unable to perform it in many cases. These facts will explain why even in the United States, according to a 1988 publication (Maguire 1988), 97% of cytologists would document their findings by description, but 72% would also enter the usual Papanicolaou groups as well.

In the Federal Republic of Germany, prolonged discussions resulted in 1975 in what was called the Munich Scheme (Table 1). The essential point was a reference of cytologic findings to the expected histologic findings, and their overall grouping in therapy-relevant categories. This grouping was no longer congruent with the old Papanicolaou groups, although the terms used for classification were not completely dissimilar.

All findings deviating from the norm should be documented using a verbal description, but an additional classification of groups should be maintained. The compromise appeared indicated because the forms of the German Cancer Detection Program required the obligatory "ticking" of appropriate boxes, with only a small space for individual remarks.

In a few years the Munich Scheme of 1975 has led to a uniform recording of cytology findings which has helped to establish a basic understanding between cytologists, gynecologists, and histopathologists.

On the international level, the discussion of recording problems in gynecologic cytology has been reopened by the end of 1988 when the National Cancer Institute published its Bethesda system.

The Munich Nomenclature II (Appendix 1) presented here supplements and differentiates the 1975 scheme, comprising also the recent international developments, and the proposals of the Bethesda system in particular; it may be explained as follows.

Appendix 1

Munich Nomenclature II

Supplement and detailed update of the previous scheme of 1975.
Verbal description and/or evaluation of all cytologic findings obligatory.
Diagnostic grouping (D) provides classification, statistical documentation, and quality control.

A Smear quality
1. Adequate
2. Approximately sufficient
3. Insufficient

in 2. and 3., the cause of inadequacy must be mentioned.

Examples of possible causes:
- Poor material (amount of cells too small)
- Inadequate fixation
- Severe degenerative cell changes
- Severe inflammation
- Heavily bloody smear
- Strong cell overlapping
- No endocervical cells

B Proliferation grading (formulated according to A. Schmitt)

C Microorganisms
- Döderlein flora with or without cytolysis
- Mixed bacterial flora
- Coccal flora/gardnerella
- Mycotic flora
- Trichomonads
- Other organisms

D Classification of cytologic findings

Group	Definition
I	Normal cell pattern corresponding to age, including mild inflammatory and degenerative changes and also bacterial cytolysis
II	Distinctly inflammatory changes in cells of squamous and cervical epithelium: cells from regenerating epithelia, immature metaplastic cells, increasingly degenerative cellular alterations, para- and hyperkeratotic cells – normal endometrial cells even after menopause – specific cell patterns such as follicular cervicitis, cellular changes in IUP, signs of HPV infection without marked nuclear changes, signs of viral herpes or cytomegalic infection

Group	Definition
	Recommendation. Cytologic monitoring if necessary, intervals chosen according to the clinical findings – if required, following therapy for inflammation or clarification by hormone administration
IIID	Cells of mild to moderate dysplasia (signs of HPV infection should be mentioned separately)
	Recommendation. Check up after 3 months
IVa	Cells of severe dysplasia or carcinoma in situ (signs of HPV infection should be mentioned separately)
	Recommendation. Histological clarification, in exceptional cases cytologic monitoring
IVb	Cells of severe dysplasia or carcinoma in situ; cells of invasive carcinoma cannot be excluded
	Recommendation. Histologic clarification
V	Cells of a malignant tumor: – Cells of squamous carcinoma (keratinizing/non-keratinizing) – Cells of adenocarcinoma, preferably with indication of origin, i.e., endometrial, endocervical, or extrauterine – Cells of other malignant tumors
	Recommendation. Histologic clarification
III	Indistinct findings: – Severe inflammatory, degenerative, or iatrogenic alterations inadequate for safe definition of benign or malignant nature – Conspicuous cells of glandular epithelium, origin from carcinoma not to be excluded; indication of endometrial, endocervical, or extrauterine nature requested, if possible.
	Recommendation. According to clinical findings, cytologic monitoring at short intervals, or immediate histologic clarification.

First, all cytologic findings are recorded as a verbal description. Positive and doubtful findings should be recorded extensively, negative findings briefly. The cytologic record represents a professional medical performance. The mere mention of a group is not sufficient to record a morphologic finding. The vast majority of experts from German-speaking countries have agreed that additional classification of findings into diagnostic groups would be indispensable for the German Program for the Early Detection and Prevention of Cancer. These groups are necessary for the classification of findings, statistical documentation, and quality control.

Secondly, assessing the technical quality of a smear is an essential part of each cytologic record, i.e., a statement of whether the cell material on the slide is sufficient, barely sufficient, or insufficient for evaluation. For slides that are only poorly evaluable or plainly insufficient, the cause has to be mentioned in the record. Such causes may be: inadequate quantity of cell material, insufficient fixation, severe inflammation, severe degenerative changes, heavily bloody smears, severe overlapping of cells, absence of endocervical cells. Any absence of endocervical cells must be mentioned by the cytologist in his/her report to inform the sender about possibly inadequate sampling or smear technique. However, I do not believe that a smear without endocervical cells should be repeated; the decision remains with the referring gynecologist.

Thirdly, according to Schmitt (1953), the grading of proliferation can be given in steps 1 to 4: 1 meaning parabasal cells; 2, smaller; 3, larger intermediate cells; and 4, superficial cells. This is a simple but meaningful way to visualize the structure and height of the vaginal epithelium.

Fourthly, microbiologic findings, as far as discernible from Papanicolaou-stained smears, should be included in the verbal description, e.g., Döderlein rods with bacterial cytolysis, inflammatory cell type with coccal flora, cellular and nuclear inflammatory changes in trichomonadal infection, etc.

Fifthly, the classification of findings in groups ; since it generally follows the Munich scheme of 1975, the comments can be fairly brief.

The group I "normal" should apply only to those findings that require no extra check beyond normal yearly prophylactic examination.

Group II with "marked inflammatory, regenerative, or degenerative changes" has been supplemented by the registration of specific cellular pictures, such as follicular cervicitis, cellular changes by IUP, signs of HPV infection without marked nuclear changes, signs of herpes or cytomegalic infection, etc. No special subgroup IIW (repeat) or IIK (control) is foreseen in the Munich Nomenclature II.

The classification of intraepithelial cervical neoplasia was left unaltered. Group IIID comprises light and moderate dysplasia, that is, changes requiring an initially short-term cytologic and colposcopic check up.

Group IVA comprises severe dysplasia and carcinoma in situ for which histologic verification is deemed necessary. Any sign of HPV infection (collocytes and dyskeratocytes) must be mentioned in both groups. Group IVb comprises the groups where severe dysplasia or CIS is present, but also cells that might derive from an invasive carcinoma – in fact, findings permitting no clear cytologic decision between intraepithelial neoplasia and true carcinoma. The category represents no specific cytologic findings of microcarcinoma, although such findings are known to be found in that group rather often.

Table 2 compares the classification of "intraepithelial neoplasia" in different nomenclature systems: the left column gives the sequence light, moderate, severe dysplasia, carcinoma in situ. The next column parallels it as CIN I, II, III, light dysplasia corresponding to CIN I, moderate dysplasia to CIN II. In CIN III, severe dysplasia and carcinoma in situ are taken together. The third column gives the Munich Nomenclature II in two groups: light to moderate dysplasia = III D, and severe dysplasia or carcinoma in situ = IVA, both identical with the previous Munich scheme of 1975. The fourth column represents the Bethesda classification: there are also two groups, but the borderline between light and severe changes is somewhat shifted towards the lighter changes. This recommendation was not adopted in the Munich Nomenclature II. Also, the Bethesda system has

Table 2. Concordance of different histologic and cytologic classifications (nomenclatures) for intraepithelial neoplasms of the uterine cervix (Soost and Baur 1990)

<table>
<tr><th>WHO</th><th>CIN</th><th>Munich II</th><th>Bethesda</th></tr>
<tr><td>Mild dysplasia</td><td>CIN I</td><td rowspan="2">Mild to moderate dysplasia (IIID)</td><td>Low-grade squamous intraepithelial lesion</td></tr>
<tr><td>Moderate dysplasia</td><td>CIN II</td><td rowspan="2">High-grade squamous intraepithelial lesion</td></tr>
<tr><td>Severe dysplasia
Carcinoma in situ</td><td>CIN III</td><td>Severe dysplasia or carcinoma in situ</td></tr>
</table>

abandoned "cervical intraepithelial neoplasia" in favor of "low-grade" and "high-grade squamous intraepithelial lesions," which I feel is a most reasonable decision.

All findings of malignant tumors, from any origin whatsoever, are comprised in group V. The detailed verbal description should specify whether it is a squamous carcinoma, keratinizing or not; a glandular carcinoma of endometrial, cervical or extrauterine origin; or any other malignancy.

Group III is meant to cover all unclear findings permitting no clear cytologic decision between benign and malignant. These are mainly found in two types of cell pictures: there are smears with severe inflammatory, degenerative, or iatrogenic cellular alteration, predominantly originating from squamous or metaplastic epithelia, but also deriving from endocervical glandular cells that are hard to distinguish from poorly preserved atypical cells. The second type is seen in smears with conspicuous glandular cells, often poorly preserved, for which an origin from endometrial carcinoma cannot be excluded. It had been proposed to collect these findings in a specific Group IIIG (glandular), but the majority of pathologist members of our society were not ready to agree.

The important point seems to be that we must have a special category for the classification of unclear cytologic findings. Even the most experienced cytologist evaluating technically perfect smears will sometimes be unable to pronounce a definitive diagnosis. If we miss this "undecided" group, we risk reports and false-positive or false-negative II and IVB, i.e., of over- or underdiagnosed cases respectively.

The Munich Nomenclature II presented here was definitively adopted and approved in December 1989 by the Executive Board of the German Society of Cytology, with positive recommendation to all its members. This system for the recording of cytologic findings can be carried out in the conventional way on routine forms by verbal description and ticking off group boxes, or by adopting a suitable code for computed printing as proposed by the Bethesda system.

References

Igel H (1947) Verhandlungen des Berliner Gynäkologen-Kongreß Oktober 1947. Zentralbl Gynakol 1369

Maguire, NC (1988) Current use of the Papanicolaou class system in gynecologic cytology. Diagn Cytopathol 4:169

Papanicolaou GN, Traut HF (1943) Diagnose of uterine cancer by the vaginal smear. Commonwealth Fund, New York

Riotton G, Christopherson WM, Lunt R (1973) Cytology of the female genital tract. World Health Organisation, Genf

Schmitt A (1953) Eine Gradeinteilung für die funktionelle Zytodiagnostik in der Gynäkologie. Geburtshilfe Frauenheilkd 13:593

Soost HJ, Baur S (1990) Gynäkologische Zytodiagnostik, 5th edn. Thieme, Stuttgart

The Bethesda System: The European Perspective

V. Schneider

Department of Obstetrics and Gynecology, The University of Ulm, Prittwitzstr. 43, 89075 Ulm, Fed. Rep. of Germany

On April 29 and 30, 1991, the Second Conference on the Bethesda System was held at the Lister Hill Center at the National Institute of Health (NIH) in Bethesda, Maryland. The meeting was sponsored by the Early Detection Branch and the Cytopathology Section of the National Cancer Institute of the United States. The purpose of the meeting was a critical evaluation of the new classification system for cervical cytological diagnosis proposed at the first Bethesda meeting on December 12 and 13, 1988. At the second meeting, 182 invited cytopathologists, cytotechnologists, and gynecologists discussed approximately 65 presentations from a variety of laboratories on their practical experience with the Bethesda System. The vast majority of the participants were from the United States and Canada.

Background

The 1988 Bethesda System is an attempt to standardize cytologic terminology. Particularly in the United States, there has been a wide variety of terminologies in use, for example, the Papanicolaou classification with various meanings attached to it, the dysplasia – carcinoma in situ terminology with various gradings and the cervical intraepithelial neoplasia (CIN) grades I–III. The introduction of the Bethesda system may also be seen as a defensive measure against mounting public criticism by the media and general public in the United States on a perceived lack of quality control in cytology. Over the last 5 years there have been reports in newspapers and journals, on television and radio about the approximately 20% false-negative rate in cervical cytology. There have been a number of well-publicized court cases with awards of hundreds of thousands of dollars against cytology laboratories who are alleged to have misreported smears from cases of cervical cancer. As a result of this public concern, Congress in Washington has introduced very strict legislation regarding laboratory activities in the

United States. The new law, CLIA 88, is now being introduced into practice. The Bethesda System should be viewed against this background as an attempt by the cytopathology profession to improve communication between cytopathologist and gynecologist.

At the first Bethesda Meeting, the following three guidelines were unanimously approved by the participants:

1. The cytopathology report is a medical consultation.
2. The Papanicolaou classification of reporting consultations is not acceptable in the modern practice of diagnostic cytopathology.
3. The Bethesda System should serve as a guideline for cytopathology reports of cervical/vaginal specimens.

There were four major deficiencies of the Papanicolaou classification cited as reasons to abolish the entire system:

1. It does not reflect current understanding of cervical neoplasia.
2. The Papanicolaou classes have no equivalent in diagnostic histopathologic terminology.
3. The Papanicolaou classes do not provide for noncancerous diagnoses.
4. As a result of numerous modifications, the specific Papanicolaou classes no longer reflect uniform diagnostic interpretations.

Second Bethesda Conference

At the second Bethesda meeting there was overwhelming support for this terminology system by almost all the participants. Of the new features of the Bethesda System, the mandatory statement on the adequacy of the specimen with the grouping of specimens as adequate, or inadequate was unanimously praised by all the cytopathologists and cytotechnologists. There were, however, considerable objections by the participating gynecologists to the term "less than optimal" because of possible legal implications. Some cytologists also felt the term difficult to define. The second essential feature of the new Bethesda System, the reduction of the number of preinvasive groups from three to two, low-grade and high-grade squamous intraepithelial lesion (SIL), was also supported by nearly all the participants. In view of the proven interobserver variability in the interpretation of intraepithelial lesions and the lack of therapeutic consequences of a more detailed system, the reduction to two groups was supported by the gynecologists as well. Finally, there were objections by the participating gynecologists, representing their major professional groups in the United States, to the statement that the cytopathology report is a medical consultation. Again they were concerned about the possible medicolegal imlications.

Most importantly, it appeared that the big commercial laboratories in the United States, some of them processing more than 1 million smears per year, were in favor of the Bethesda System. The combined support of academic cytopathology and commercial laboratories provides considerable support for this new classification system in the United States. After a well-organized and highly informative 2-day meeting, it was concluded that minor modifications will be worked out by an editorial committee and will soon be published. The few European participants at this meeting represented Eastern Euorpe, the United Kingdom, Germany, Italy, The Netherlands, and Sweden. They reported on a rather unenthusiastic reception of the Bethesda System in Europe. In several European countries, standardized terminologies are established, so the need for a new system is less prevalent. Time will tell which of the competing terminology systems will ultimately prevail.

III. Benign Changes in Cytology

Fluorescence Microscopy for Pneumocystis carinii

K. Wehle, T. Küpper, and P. Pfitzer

Institute of Cytopathology Moorenstrasse 5, 40225 Düsseldosf, Fed. Rep. of Germany

Introduction

Diagnoses of benign diseases are rare in diagnostic cytology and of less importance in daily routine than the proof of malignant cells. However, there was a recent change in lung cytology concerning the diagnosis of *Pneumocystis carinii*, made possible by fluorescence microscopy.

Before 1981, *Pneumocystis carinii* pneumonia was a rare disease, first described in dystrophic children and mainly seen in oncologic or transplant patients receiving immunosuppressive treatment. In recent years, it has become the most important opportunistic infection in patients suffering from AIDS. Occurring in at least 75% of these patients and leading to death in about 25%, it is their major initial presentation and most significant immediate cause of death.

Since clinical and radiological features are not specific and no reliable serological or culture methods are yet available, the diagnosis of *Pneumocystis carinii* pneumonia still depends on the morphological demonstration of the pneumocysts. For this, the cytological evaluation of bronchoalveolar lavage samples with a mean sensitivity of 93% (14 series; Wehle et al. 1991) has been shown to be superior to histology of transbronchial lung biopsies with a mean sensitivity of 84% (12 series; Wehle et al. 1991).

Though its fungal or protozoal nature has not been finally elucidated yet, an extracellular, usually intra-alveolar life cycle has been claimed (Gutierrez 1989). Three developmental stages have been described: cysts, intracystic bodies or sporozoites, and trophozoites – all stages demonstrating different staining properties (Bedrossian et al. 1989; Pintozzi et al. 1979). With accelerated development of the parasite, trophozoites and intact and ruptured cysts accumulate to the characteristic alveolar casts. For the demonstration of *Pneumocystis carinii* in bronchoalveolar lavage samples, a variety of histochemical stains have been recommended. According to their different affinity to the developmental stages of *Pneumocystis carinii*, they are

Recent Results in Cancer Research, Vol. 133

classified into three groups. In Papanicolaou-stained preparations *Pneumocystis carinii* can usually be detected in the form of the characteristic alveolar casts. Sometimes cyst walls and intracystic bodies can also be discerned (Bedrossian et al. 1989; Wehle et al. 1991). The methenamine silver stains according to Gomori and Grocott and toluidine blue O selectively stain the cyst wall structures. The Romanovsky stains permit the identification of trophozoites, though their interpretation requires considerable experience on the part of the examiner (Bedrossian et al. 1989). For rapid diagnosis, phase-contrast or dark-field microscopy on fresh unfixed samples have been recommended.

Recently, several monoclonal antibodies have become available for immunochemical demonstration of the parasite (Linder and Radio 1989). For the final cytological diagnosis of *Pneumocystis carinii* in bronchoalveolar lavage samples Bedrossian et al. (1989) demanded a combination of stains: screening for alveolar casts in Papanicolaou-stained slides should be confirmed by a silver stain for the demonstration of cysts.

In 1984, Ghali et al. reported a specific morphological pattern of *Pneumocystis carinii* cysts in Papanicolaou-stained preparations viewed under fluorescent light. According to their publication, we started a retrospective and prospective study on the efficiency of applying this method on routine cytology specimens.

Materials and Methods

Between 1986 and 1989 we received 213 bronchoalveolar lavage samples of 145 patients, clinically suspected of having *Pneumocystis carinii* pneumonia. A total of 98 patients were suffering from AIDS, eight from acute leukemias, and four from malignant lymphomas. A total of 27 patients were receiving immunosuppressive treatment after organ transplantation, and eight were treated with corticosteroids for other reasons.

Three slides of each lavage sample were prepared, in 129 cases using cytocentrifugation (Cytospin 2, Shandon) and in 84 cases as smears. Staining was carried out according to the standard Papanicolaou procedure. All slides were screened by a cytotechnician and a cytopathologist in routine light microscopy and afterwards examined in ultraviolet light using an epi-illumination fluorescence microscope (Zeiss IVFI, filters FT 510, LP 520) at 630-fold magnification.

In 43 cases additional slides of the lavage samples were available for Grocott's methenamine silver stain. In 77 cases transbronchial lung biopsies from the same bronchoscopy were submitted. In 145 cases the results of the microbiological examination were available. In all cases the morphological results were compared with the clinical history. Cytospin preparations of 84 lavage samples from 76 patients without known immunological deficiency served as controls.

Morphology

Pneumocystis carinii presented in light microscopy of Papanicolaou-stained lavage samples in the form of the typical foamy alveolar casts with central eosinophilia and peripheral basophilia. In most of the positive cases, parts of alveolar casts broken by the lavage or preparation procedures were found. Occasionally two to eight eosinophilic intracystic bodies could be identified.

Fluorescence microscopy equally revealed complete and disrupted alveolar casts. Single pneumocysts appeared as circular, crescentic, or cup-shaped structures of 4–6 μm in diameter, emitting an intensive greenish fluorescence. Many cysts contained two mirror-image reniform or parenthesis-like bodies. Correlation between these reniform bodies and the inclusions seen on light microscopy representing the sporozoites could never be observed.

By their brilliant fluorescence and characteristic morphological pattern, even very few singular cysts could be detected despite obscuring cellular material such as macrophages, ciliated cells, or erythrocytes. In smears covered with abundant mucus, fluorescence was usually faint, therefore we prefer cytospin preparations. However, in all positive cases investigated, at least some pneumocysts could be clearly identified by fluorescence.

In 78 of the 79 cases positive for *Pneumocystis carinii* we observed round or oval inclusions within the alveolar macrophages measuring 1–3 μm in diameter and emitting a brilliant green fluorescence. By tracing stages from intact recently phagocytosed cysts to very small granular particles, sometimes within the same macrophage, we could identify them as degradation products of pneumocysts. These inclusions could not be observed in only one case with very few and ill-preserved macrophages. By light microscopy, the macrophage inclusions were never identified. In some of the Grocott-stained lavage samples we found intracellular deposits of the same size and configuration although they were difficult to distinguish from other deposits such as smokers' pigment.

Fluorescence of casts, cysts, and degradation inclusions was well preserved even in 4-year-old slides, though for light microscopy much of the staining had been lost. In the alveolar macrophages of 67 of the 84 control samples various intracytoplasmatic fluorescent inclusions were found. However, their irregular size and outline as well as the yellow or reddish fluorescence could be easily differentiated from the degradation inclusions of *Pneumocystis carinii*. Furthermore, they often appeared brown on light microscopy. Neither the fluorescence pattern of the intact pneumocysts nor the degradation inclusions could be observed in any of the control samples.

Diagnostic Reliability

Using the methods described, *Pneumocystis carinii* was demonstrated in 79 of 213 bronchoalveolar lavage samples from 145 patients, i.e., 37% of all episodes clinically suspicious for *Pneumocystis carinii* pneumonia. By light microscopy of the Papanicolaou-stained samples already 70 of 79 positive cases, i.e., 88.6%, could be correctly diagnosed – by the presence of complete foamy alveolar casts in 62% and of disrupted alveolar casts in 88.6%. Additionally, fluorescence microscopy revealed the presence of single fluorescent cysts in 88.6%, of intact phagocytosed cysts within the alveolar macrophages in 68.4%, and of degradation inclusions in 98.7%. By applying these inclusions as an additional criterion for the diagnosis, all cases of infection could be correctly diagnosed by fluorescence microscopy, i.e., a sensitivity of 100%. In four patients, three of whom were already receiving treatment, macrophages with fluorescent degradation inclusions were the only criterion suggesting the presence of *Pneumocystis carinii*. In these four cases the clinical history, with rapid improvement after specific chemotherapy, supported our diagnosis.

In the 134 lavage samples negative for *Pneumocystis carinii* by all morphological methods, the clinical course did not give any evidence that a false-negative diagnosis might have been made. As there were no false positives either, specifity of all morphological methods was 100%.

Discussion

For the cytological diagnosis of *Pneumocystis carinii*, Bedrossian et al. (1989) and Naryshkin et al. (1991) demand a Papanicolaou stain to be screened for foamy alveolar casts for rapid diagnosis, combined with a special stain for *Pneumocystis carinii* cyst walls, i.e., a silver stain, for confirmation, at least in the cases negative for foamy alveolar casts. Additionally, a Gram-Weigert stain may be useful for the intracellular demonstration of intact microorganisms (Bedrossian et al. 1989). Only this combination of at least two stains would result in an optimized cytological sensitivity.

Taking foamy alveolar casts as the only criterion for the presence of *Pneumocystis carinii* may result in a sensitivity as low as 58% (Chandra et al. 1988) or 62% for our series. Fragments of alveolar casts disrupted by lavage or preparation procedures as an additional light-microscopic criterion improved our sensitivity to 89%.

Fluorescence microscopy of Papanicolaou-stained bronchoalveolar lavage samples permits a specific identification of *Pneumocystis carinii* cysts in the form of circular, sometimes crescent- or cup-shaped structures of 4–6 μm in diameter and bright greenish fluorescence (Pfitzer et al. 1989; Wehle et al. 1991). The presence of two mirror-image reniform bodies in many cysts makes this fluorescence pattern closely resemble to the morphology of the parasite in the cyst-specific silver stains, where the reniform bodies were shown to be cyst wall structures by electron microscopy (Watts and Chandler 1985). Fluorescence permits the identification of the cysts even when obscured by cellular material or when phagocytosed by alveolar macrophages. Furthermore, we found fluorescent inclusions within the alveolar macrophages in 98.7% of cases, which could be shown to be degradation products of pneumocysts by tracing all steps from intact phagocytosed cysts to very small granular particles. These inclusions were the only criterion permitting the diagnosis of *Pneumocystis carinii* in four of 79 cases, thus enhancing the sensitivity from 95% to 100% for our series.

In conclusion, fluorescence microscopy of Papanicolaou-stained bronchoalveolar lavage samples permits a highly sensitive and specific diagnosis of *Pneumocystis carinii*, even if only few organisms are present. No additional time-consuming preparation or staining procedures, which would be necessary for the combination of stains demanded by Bedrossian et al. (1989) are required. The screening for foamy alveolar casts and their fragments in light microscopy is confirmed by the demonstration of the specific fluorescence pattern of the cyst wall structures. Diagnostic sensitivity is even enhanced by the identification of phagocytosed intact cysts and their degradation products within the alveolar macrophages.

References

Bedrossian CWM, Mason MR, Gupta PK (1989) Rapid cytologic diagnosis of Pneumocystis: a comparison of effective techniques. Semin Diagn Pathol 6: 245–261

Chandra P, Delaney MD, Tuazon CU (1988) Role of special stains in the diagnosis of Pneumocystis carinii infection from bronchial washings in patients with the AIDS. Acta Cytol 32:105–108

Ghali VS, Garcia RL, Skolom J (1984) Fluorescence of Pneumocystis carinii in Papanicolaou smears. Hum Pathol 15:907–909

Gutierrez Y (1989) The biology of Pneumocystis carinii. Semin Diagn Pathol 6: 203–211

Linder J, Radio SJ (1989) Immunohistochemistry of Pneumocystis carinii. Semin Diagn Pathol 6:238–244

Naryshkin S, Daniels J, Freno E, Cunningham L (1991) Cytology of treated and minimal Pneumocystis carinii pneumonia and a pitfall of the Grocott methenamine silver stain. Diagn Cytopathol 7:41–47

Pfitzer P, Wehle K, Blanke M, Bürrig KF (1989) Fluorescence microscopy of Papanicolaou stained broncho-alveolar lavage specimens in the diagnosis of Pneumocystis carinii. Acta Cytol 33:557–559

Pintozzi RL, Blecka LJ, Nanos S (1979) The morphologic identification of Pneumocystis carinii. Acta Cytol 29:727–731

Watts JC, Chandler FW (1985) Pneumocystis carinii pneumonitis – the nature and diagnostic significance of the methenamine silver-positive "intracystic bodies". Am J Surg Pathol 9:744–751

Wehle K, Blanke M, Koenig G, Pfitzer P (1991) The cytological diagnosis of Pneumocystis carinii by fluorescence microscopy of Papanicolaou stained bronchoalveolar lavage specimens. Cytopathology 2:113–120

IV. Quality Control

Quality Control in Cytology

O.A.N. Husain

Charing Cross Hospital, London, UK

The matter of quality control does require two major approaches:

1. *Internal quality control*: this is the practice within a laboratory and demands a distinct operation under the control of a senior member of staff.
2. *External quality assurance*: this is usually established at an interlaboratory level and run by a professional peer group, such as an institute or college of pathology, and in any event not by an employer so as to obtain a fair assessment of the practice in each laboratory from its individual members of staff.

Perhaps before we consider each item it is essential to establish the structure of practice in a country and in the 1970s we in the United Kingdom, as officers of the British Society for Clinical Cytology (BSCC), laid down a code of practice relating to both staff and institutions and to assemble the levels of training and qualifications, continuing education and assessment within a department. Particular consideration was given to the facilities and atmosphere of working conditions within such a place and this included such essential items as suitable microscopes, adequate space, work flow areas and ergonomic positioning at bench and microscopes with optimal rest schedules and controlled outputs per worker.

All this was enacted, submitted and passed through the Royal College of Pathologists to the Department of Health to ultimately come out in the form of a report which also included a code of clinical practice. This took the form of a *Report of the Inter-collegiate Working Party on Cervical Cytology Screening* (1987). The bodies involved were the Royal College of Obstetricians and Gynaecologists, the Royal College of Pathologists, the Royal College of General Practioners and the Faculty of Community Medicine. This is strongly recommended reading and is obtainable from the Royal College of Obstetricians and Gynaecologists, 27 Sussex Place, Regent's Park, London NW1 4RG.

Recent Results in Cancer Research, Vol. 133

Training and Qualifications. The very basis of quality practice is a proper formal, didactic and practical training programme for each of the workers in the laboratory. For the pathologist there is an increasing exposure to cytological preparation and interpretation, still probably insufficient, but it is to the technical staff both senior and junior and especially to the cyto-screener, who is the main bulwark of the safety net of good practice, that we should direct our attention. It is here that a comprehensive training system should exist. This has been achieved to a substantial extent in the United States but in Europe it is still far from satisfactory. The analysis in 1976 by the European Federation of Cytology Societies was disappointing in that a core curriculum could not be agreed amongst national societies for any substantial uniform training and practice. We in the United Kingdom have only achieved short 6-week intensive courses at five to eight training schools, although the number is increasing. Such courses are combined with substantial in-service experience but it is by no means sufficient to establish a high-quality cadre of skilled workers and it is on this basis that the new European Community initiative of attempting uniformity of training programmes throughout European countries is welcomed (Douglas 1992).

Internal Quality Control

This includes monitoring of the whole exercise from collection of the specimen from a patient to the ultimate reading of the report by the collector of the sample. It is possible for all of this to be monitored from the laboratory.

The baseline of good practice in a laboratory is its so-called standing order of practice which should be exposed for all to see and a complete laboratory manual of techniques which must be dated for each procedure and updated when necessary. Individual job descriptions per worker are also mandatory and these are summarised in Appendix 1.

The potential hazards to be monitored can be listed as follows: identification errors, sampling, processing, screening, reporting, histological and clinical correlation, and storage and retrieval.

Identification Errors

Identification errors are summarised in Table 1.

Sampling

The hazards in this category can be divided into:

- A difficult patient (obese)
- An inexperienced sampler

- A poor instrument
- Poor spreading and fixing of smear
  (Too thick/thin/too much blood/pus, etc.)

It is surprising how many sampling errors external to the laboratory can be monitored by the laboratory. It has been shown that even in a well-run programme a false-negative rate of around 20% can occur and that the sampler rather than the sampling instrument can be most at fault (van de Graf et al. 1987).

The poor collecting instrument has been blamed by many and an improved shape of the head of the spatula has resulted in about a 20% improvement in sampling. In the United Kingdom this followed much research by members of the BSCC over a 10-year period which resulted ultimately in the Aylesbury spatula (Wolfendale et al. 1987). This was not the whole story as a number of other spatulae – some with movable heads, the Cervex (a plastic broom-like instrument) and the endocervical brush – are all contributing to improved sampling.

Table 1. Identification errors

Cause	Responsibility
Wrong slide labelled in clinic	Clinic doctor/nurse
Poor legibility on slide or form	Clinic doctor/nurse
Incorrect slide labelling after processing	Laboratory processor
Reading the wrong smear from tray	Screener
Inadequate report from laboratory	Screener/chief technologist or pathologist
Transcription error in report	Secretary
Miss mailing of report	Clerk
Non-arrival at destination	Internal/External Mail/porters

Processing

Processing has its own problems through poor schedules or work flow patterns, but it is the quality of staining and the dangers of cross-contamination that create the real hazard with a potential transfer of cells from slide to slide of 15%–100% (Husain et al. 1978). There is a serious need to create a potentially non-contaminating single-slide stainer but all my efforts for the past 20 years to get one manufactured have resulted only in bigger bulk stainers.

Screening

Here the basic training and maintenance of competence, the work pattern, the laboratory layout, the atmosphere, the peaceful milieu to work in, and senior technical and medical monitoring come into full play. It takes very little to realise that pressure of work, patchy instruction, lack of properly laid down codes of reporting and referral for second opinion or double screening will result in errors or drifts (where the same cytologist alters his/her diagnosis on different occasions).

The screener's primary task is to exclude any evidence of disease processes, especially neoplastic, and it has become the custom to record freedom from specific infections and hormone patterns as well as neoplasia and certainly to make a comment if the cell pattern is out of context with the woman's present physiological state or menstrual cycle.

In our laboratory we have used a rough scoring of numbers of squamous cells in 10000s on a range of C1–C4 when C1 indicates 10000 cells and C4 40000 or more. A similar type of score is given to endocervical columnar cells (10–60+) as E1–E4. A further code for the obscuring of cellular content by blood (B1: a quarter of the smear obscured; B4: completely obscured) and similarly with pus cells P1–P4 can be used. The rapid signing out of such a form may then permit a degree of confidence, having identified the cytotechnologist and assessed the quality of the report, without needing to look at the smear or, alternatively, to check when in doubt. It is at this level that most errors occur in cytology and, as the laboratory is now being sued for mistakes, a proper code of conduct requires that this degree of cross-check be practised.

The matter of doing a random 10% check on all negative smears is considered as an adequate safeguard but I must refer to the paper by Melamed (1973). He calculated that to achieve a 95% chance of proving a cytoscreener was in 10% error when screening some 10000 smears a year it would take about 13.2 years if there was a 5 per 1000 abnormality rate in an unscreened population but, if it is a re-screened population with an incidenece rate of less than 2 per 1000, it would take 66 years to prove it – an obviously unrealistic task.

We ourselves have the initials of every person, screener, chief technologist or cytopathologist, who sees the smear signed on the front of the form and find that somewhere between 15%–20% have two sets of initials and almost 10% have three. These are, however, on selected high-risk categories of patients such as previous abnormal smear cases, those with symptoms and signs, most if not all postmenopausal cases and those from

Table 2. Christie Hospital, Manchester: re-screen of negative smears (n = 14437)

Review of first smear	n	Rate/1000
Negative	16	1.11
Positive	9	0.60
	25	1.71

Table 3. Christie Hospital, Manchester: missed positives considered as percentage of true positives

	n	Rate/1000
Positive at first screening	143	9.94
"Missed" positives	25	1.71
"True" positive	168	11.65
Error rate	25/168	14.9%

special urogenital medicine clinics. The price of good-quality screening is high but the price of mistakes is exorbitant and results in tragedy.

Published records of error rates of 5%–30% and more exist in the literature but one analysis carried out many years ago by Dr. R. Yule in Manchester showed that, when he invited back some 14000 women who had had a negative smear 3 months previously, he found some 25 more positives in that group and by re-screening their previous smears around ten had been missed in the previous smears while the remainder were probably sampling errors, assuming that no new disease had occurred in 3 months (Tables 2, 3; Husain et al. 1974).

In short, this finding has been repeated many times since then to prove that such an error rate of 15% is not uncommon and is composed predominantly of a "sampling error" but that a "detection error" is still substantial.

Reporting

I put reporting separately as it is essential that standardisation of reports is a most essential feature of accurate communication of what the report means and that such reporting is consistent with other laboratories and is accepted professionally. I always teach that there are three parts to a cytology report:

Description. This is what one sees and is inevitably written in cytological terms.

Interpretation. This is given in histological terms of what the smear suggests as existing in the cervix.

Action. Advice is given on the basis of what is seen in the smear and what the cytopathologist would recommend as the next move. This is entirely focussed on investigation and not treatment as that is the prerogative of the clinician, but the advice code ranges from "normal re-call" to urgent referral for colposcopy and/or urgent biopsy.

The use nowadays of the Bethesda Terminology (1989) is one way of categorising the lesion down to small detail and, though it is elaborate, it caters for every eventuality and is eminently computerisable while allowing for free text where necessary (Schneider 1992). Other more simple terminologies exist but these must be accepted professionally between the clinical client and the laboratory to avoid misinterpretation of reports.

Histological and Clinical Correlation

Histological correlation is relatively easy to enact as it is mandatory for good quality control to compare any histology resulting from a cytologically diagnosed or detected case. Not all get biopsied and some with only small punch biopsies which may well be less reliable than the cytology. The same occurs for colposcopy which must be used to correlate and/or confirm the cytology, but again a small remote lesion may well be missed by the colposcope. It is said that each of the three methods, cytology, histology and colposcopy, has a similar error rate for different reasons and that by "correlation" it is not always possible to say which is correct.

Table 4 shows the correlation of histology to cytology in St. Mary's Hospital, Manchester, where recognised experts were practising (Husain et al. 1974). The table has been slightly modified in that the cytological diagnoses are given here in terms of the Bethesda system.

Table 5 shows the effect of cutting further sections from the original blocks of a cone biopsy specimen in the Vancouver survey which demonstrated some 17% increase in severity of the lesion. No one is perfect and perfection increases the workload and cost (Nichols et al. 1968). An average of 12 blocks was taken from each cone and on average 90 additional sections were examined.

Storage and Retrieval

The matter of length of storage time of smears and reports has been the subject of debate over the years. It was once said, and practiced in America,

Table 4. Comparison of histology/cytology reports: St. Mary's Hospital, Manchester, 1970–1972

Histology	*n*	Squamous carcinoma	HG SIL	LG SIL	Atypical	Negative
Squamous carcinoma	78	44	31	1	1	1
Carcinoma in situ	91	3	71	12	1	4
Dysplasia	126	1	54	59	5	7
Benign	105	0	7	34	64	–
	400	48	163	106	71	12

HG SIL, high-grade squamous intraepithelial lesion; LG SIL, low-grade SIL.

Table 5. Difference in final diagnosis between single and step serial sections of serially blocked cone biopsies

	Cones with one section from each block	Cones with step serial sections from each block		More information
		Finding	*n*	
Negative	38	Negative	23	
		Dysplasia	8	15
		In situ	7	
Dysplasia	133	Dysplasia	79	
		In situ	54	54
Carcinoma in situ	422	In situ	392	
		Micro-foci	24	30
		Occult	6	
Micro-invasive foci	18	Micro-foci	15	
		Occult	3	3
Occult invasive carcinoma	13	Occult	13	0
Lymphatic permeation	0		3	3
	624			105 (17%)

that there should be two negative smears before an earlier one was discarded. It is now apparent that a much longer time of storage is necessary to confirm long-term development of a lesion especially with newer identification techniques or markers and that 20 years is about right. The logistics of this have to be considered seriously and some intermediate compromise probably effected, as space and weight of storage become a problem.

External Quality Assurance

This aspect of interlaboratory comparison to achieve a high standard overall has been pursued in many directions and the main approaches have been:

1. Exchange of smears between laboratories – if not fewer than five units are involved, it can give some idea of the laboratory's behaviour amongst its fellows with peer review on problem cases. However, it fails to test effectively the individual worker as it is probably the best opinion which comes out of each laboratory. Even so, observer variation is embarrassingly high (Evans et al. 1974; Husain et al. 1984; Thomas et al. 1988).
2. The review of previous smears by a third party, if and when a subsequent positive occurs, is a good way of checking as an internal quality control test but is too haphazard for a regular monitoring from outside.

3. The use of the inspectorate tray – or what is now considered the New York system – where an invigilator brings in trays of unknown smears to test first the cytoscreener, usually with around ten smears of mixed negative and positive cases, then the senior technical staff with part-screened and reported smears and some unscreened. A similar test series is also given to the pathologists. This does produce a very good estimate of each person's prowess and has become the chosen procedure for the United Kingdom (Collins and Patacsil 1986). The criticism here is that it is still an artificial environment and tests a person somewhat more "keyed up" (for better or worse) than would apply in routine practice. In the United Kingdom we have adopted the practice of testing at 6 or 12 monthly intervals depending on the finance available and, although not perfect, it appears to be a quite salutory way of testing staff. Our experience, though yet unpublished, does permit a degree of confidentiality and the opportunity to interview and advise the person with below-threshold results and to introduce some form of re-education especially on a one-to-one basis using a multiple-head microscope.

In fact it is our confirmed opinion that for both internal and external quality control the multihead conference microscope is the most valuable instrument in the laboratory and should be in constant use, as should television projection if available so that everyone involved in cytology is correlating cell by cell. At a recent automation conference where automated scanners were being evaluated, it was embarrassing to find our own accuracy being challenged when we firmly believe our departments are near perfect. There is still considerable room for improvement.

Appendix 1

Duties of All Grades of Staff

Duties of Clerical Staff

1. Organise and maintain a satisfactory accession register (e.g., daybook or computer data entry)
2. Maintain satisfactory file system for rapid and effective recall of data
3. Type out reports
4. File and/or despatch copies of reports
5. Maintain a comprehensive and efficient disease index
6. Prepare statistical data
7. Monitor quality control data
8. Maintain high quality of service to the user
9. Public relations

Duties of Technical Processor

1. Unpack and identify specimens, number of slides, etc., match with forms
2. Process materials if necessary to prepare smears/sections/filters, etc.
3. Stain, mount and label
4. File slides away after use
5. Maintain work-bench and machinery in good clean working order
6. Maintain adequate stocks of processing materials

Duties of a Cytotechnologist

1. Identify slides and match with request forms
2. Comprehensively screen the whole of the smear material
3. Report on adequacy of smear
4. Report on freedom from malignancy and its precursors
5. Assess as an optimal test
6. Identify other significant findings (e.g., trichomonas, atypical metaplasia, unexplainable hormone changes, cell debris, etc.)
7. Write a short descriptive report of abnormal findings and answer clinical queries
8. Receive feedback information
9. Accomplish reasonable and reliable output
10. Take part in quality control exercises
11. Keep up to date with processing and screening techniques

Duties of Chief Cytotechnologist

1. Organise and run technical side of the laboratory
2. Perfect and record processing techniques
3. Train and maintain quality of technologist staff
4. Monitor processing and screening techniques
5. Check screened material which has both normal reports and the smears for submission to cytopathologist
6. Organise and conduct feedback to cytoscreener
7. Organise and conduct quality control under guidance of cytopathologist
8. Undertake research and development

Duties of Cytopathologist

1. Establish optimum size and calibre of departmental staff, space and equipment for service required

2. Establish and maintain efficient and reliable work flow to provide a satisfactory service
3. Create formal training programmes and a variety of in-service experience with seminars and case studies to achieve a high-quality staff
4. Create and maintain a comprehensive, lucid and accurate laboratory technique manual in conjunction with the chief cytotechnologist
5. Maintain a high morale and good working conditions
6. Organise and maintain quality control procedures with the chief cytotechnologist; make periodical reports to central authority
7. Report all abnormal smears and be closely involved in a proportion of others (10%–30%)
8. Keep abreast of current advances and undertake research and development

References

Bethesda Terminology (1989) The 1988 Bethesda system for reporting cervical/vaginal cytologic diagnoses. Acta Cytol 33:567–574

Collins DN, Patacsil DP (1986) Proficiency testing in cytology in New York: analysis of a 14 year state program. Acta Cytol 30:633–642

Douglas G (1992) Report of the Manager of the European Commission Training Project on cervical cancer screening. Cytopathology 3:23–25

Evans DMD, Shelley G, Cleary B, Baldwin Y (1974) Observer variations and quality control of cytodiagnosis. J Clin Pathol 27:945–950

Husain OAN, Butler EB, Evans DMD, Macgregor JE, Yule R (1974) Quality control in cervical cytology. J Clin Pathol 27:935–944

Husain OAN, Grainger JM, Sims J (1978) Cross contamination of cytological smears by bulk automated methods. J Clin Pathol 31:63–68

Husain OAN, Butler EB, Woodford FP (1984) Combined external quality assessment of cytology and histology opinions: a pilot scheme for a cluster of five laboratories. J Clin Pathol 37:993–1001

Melamed MR (1973) Presidential address of 20th annual scientific meeting, American Society of Cytology. Acta Cytol 17:285–288

Nichols TM, Boyes DA, Fidler HK (1968) Advantages of routine step serial sectioning of cervical cone biopsies. Am J Clin Pathol 49:342–346

Report of the Intercollegiate Working Party on Cervical Cytology Screening (1987) Royal College of Obstetricians and Gynaecologists

Schneider V (1992) The Bethesda system – the European perspective: report on the second conference on the Bethesda system for reporting cervical/vaginal cytological diagnoses. Cytopathol 3:27–29

Thomas GDH, Head C, Thorogood J (1988) Quality assessment in cervical cytology: a pilot study. J Clin Pathol 41:215–219

Van de Graf Y, Vooijs GP, Gaillard HLJ, Go DMD (1987) Screening errors in cervical cytologic screening. Acta Cytol 31:434–438

Wolfendale MR, Howe-Guest R, Usherwood MM, Draper GJ (1987) Controlled trial of a new cervical spatula. Br Med J 294:33–35

V. Who is a Cytologist?

Who is a Medical Cytologist?

O.A.N. Husain

Charing Cross Hospital, London, UK

The challenging title "Who is a medical cytologist?" brings forth considerable discussion and differences in opinion and has been hotly debated in meetings and conferences over the past few years. I chaired a symposium entitled "Cytology as an Independent Speciality" at the World Association of Societies of Pathology (WASP) in Jerusalem in 1981 and then again at the European Federation of Cytology Societies in Prague in 1982 in a symposium entitled "Who should practise as a medical cytologist?". A year later it was the subject of my Goldblatt Award lecture in Montreal in 1983. Here the title really was "Who is a medical cytologist?".

Now that we are addressing the subject again, I would like to give a running list of the names of most of the main pioneers and activists in this field. They are listed alphabetically and not chronologically, and this list is not to be considered complete. They almost all have one thing in common: they were not histopathologists:

> Atay, Ayala, Babes, Blonk, Cardozo, Crepinko, Ehrlich, Figueredo, Fischnaller, Franzen, Frost, Grunze, Hauptmann, Hattori, Hayata, Ide, Jenny, Koss, de Lemos Bastos, Leonhard, Luscan, Masubuchi, Matzel, Monsus, Mouriquand, Naujoks, Papanicolaou, Pedio, Reagan, Spriggs, Stoll, Takeda, Tortora, Villaplana, Wachtel, Wagner, Webb, Wied, Wihan, Witte, Wittekind, Zajicek, Zimmer, Zinser, et al.

It does not follow that a histopathologist does not make a good cytologist as they start from a very fortunate position of knowing well the normal and abnormal macro- and microanatomy of the body tissues.

Perhaps I should mention here that in the United Kingdom we use the generic name of "pathologist" for all who work in a laboratory and this results in a range of histopathologists (sometimes termed morbid anatomists), chemical pathologists, microbiologists or pathologists, and haematologists or haematopathologists. I will use the term histopathologist in this article to identify the "tissue pathologist".

Table 1. Senior cytologists in Europe

Discipline	%
Histopathologist	35.6
Gynaecologist/surgeon	34.8
Physician/haematologist	7.6
Anatomist	0.27
Medical cytologist	8.85
Non-medical graduate	12.85
	100.00

Pre-eminent in the above list was George Papanicolaou who was in no way a histopathologist. In every country where the field developed, the pioneering work was performed by a wide range of medical generalists and specialists, and in many instances those clinicians who, seeing the value of such an approach, actually did undertake to provide the service themselves. This is very apparent in the case of gynaecology where, getting no enthusiastic response from the histopathologists, they created their own cytology screening programme for the cervix.

Table 1 shows a list of those doctors who were heading cytology services in Europe from an analysis made in 1976. This resulted from a survey conducted by the European Federation of Cytology Societies and will be fully published elsewhere.

Another prominent specialist was the haematologist who also specialised in studying cells in smears. In many ways, during World War Two and thereafter, we in the United Kingdom kept the study of cells under the combined umbrella of histopathology and haematology, and I was one of these. In some cases it was the experimental approach that fascinated us: the handling of cells by special techniques and stains, including intra-vital stains, allowing us to explore the nature of cellular change.

As time went on it was inevitable that cytology should come under the umbrella of the histopathologist if only because the cytodiagnostic process was eroding the practice of the histopathologist. In the United Kingdom we formed the British Society for Clinical Cytology; note the title – it was not "of Clinical Cytologists" but of those for whom it was practised – thus bringing into our ranks all enthusiasts who needed the practice and service and who, then in turn, became our greatest protagonists.

In the United States cervical cytology screening was a money spinner and flourished in the hands of both cyto-specialists and some histopathologists who were more enlightened. In the United Kingdom, however, it was considered unscientific and unreliable by the histopathologist as well as being a heavy and boring routine workload. In consequence it was left predominantly to those enthusiasts who largely lacked the proper grounding in histopathology but were a hard-working dedicated group of doctors and technologists who produced a substantial cadre of workers and achieved great results in both service and research.

The somewhat tardy and restrictive way in which some histopathologists accepted the responsibility of running a cytology department has resulted in a feeling that the histopathologist is not always the best person to provide the service. It was the advent of fine needle aspiration (FNA) – mainly from Scandinavia in the 1970s – that resulted in another sub-speciality of "cellular pathologists" who in the main did histopathology but even they were being separated out from their pure histopathology cousins. Even FNA was to emerge slowly in the United States and Europe, and it was the final demand by the radiologists who submitted deep needle aspiration samples to the histopathology department demanding a diagnosis that forced the histopathologist eventually to acquire those interpretative skills. On top of this came the need to collect the sample properly and provide for the various special stains needed to identify a variety of intracellular constituents.

It is against this background that we are now trying to assess what makes a good cytopathologist. Historically the pioneers are all dying off and the provision of a skilled service has to be established and succeed. Inevitably the clinician who took on the practice of interpreting the smears became the best collector of the specimens and was, in effect, the clinician dealing with the patient. For histopathologists this was unusual and their custom of "waiting for the meat to arrive" to be examined held them back from actually dealing with patients and physically collecting the samples.

In essence we have to balance two opposites. The histopathologist, with his/her wide experience of macro- and microtissue pathology, has, inevitably, a minority interest in the world of smears where the three-dimensional structures of tissue pathology do not exist but the individual cells do – often spread out in a particular way compared with histology. For example, the tubules of a glandular lesion histologically become a morular or papillary cluster in a smear. In addition, it is the finer detail under greater magnification, to study the cellular and intracellular patterns and changes, that becomes important. The average histopathologist is accustomed to using the ×4 objective rather than the ×40; the oil immersion lens is almost never used. Moreover, the use of the more refined cytochemical and immunocytochemical stains do lend themselves uniquely to cytology with its atraumatic sampling and rapid service on fresh specimens – a habit not associated with histopathology other than a rather cumbersome frozen-section service. An "in-clinic service" is an extreme example of this and in Stockholm at the Karolinska an on-the-door service has been provided now for many years by a rota of cytopathologists so that a patient could even take back the report of a diagnostic needle test to the same half-day clinic. Such is the extreme but it is beginning to be considered necessary to position such a laboratory in juxtaposition to the operating theatres, clinics and out-patient services.

Inevitably there is a division of interests from the morbid anatomist dealing mainly with post-mortems and large specimens of surgical pathology to the rapid functional service of the cytologist; hence, recruitment is different. I firmly believe that in this dynamic rapidly changing world anyone dealing with a cytological diagnostic service has to be very well grounded, trained and examined in the fundamentals of macro- and micro-anatomy before being allowed to practise cytopathology. Training should be very similar for both specialists. In fact, the ideal term should be the more generic "cellular pathologist". Practitioners could then apply themselves more to either histopathology or cytopathology as desired.

There is no space to discuss the full curriculum of such training but we are now achieving the overall agreement of semester involvement for each laboratory speciality in Europe and this should include both histology and cytology experience in substantial degree as close correlation between the two skills should be taught and practised at all times.

For the other more selective practitioners dealing with cervical cytology or even breast or skin cytology there should be a place for a sub-speciality but here again a substantial didactic and practical training course should be accomplished to cover the micro-anatomy and physiology of the body generally and a limited licence issued to conduct a practice in these restricted fields. It would be folly to deny this group of enthusiasts the opportunity of providing what would be an excellent dedicated service in these rather specialised fields.

In summary, we are now on the threshold of a consolidated form of practice of cytology and the training syllabus should be the same for all histopathologists and other laboratory or clinical specialists wishing to practise. From then on the general cellular pathologist should be in charge of the larger diagnostic cytology department and be prepared to provide a wide-ranging rapid service which may well include the sampling of the specimen.

Cytology Training in the European Community: European Community Training Programme in Cervical Cancer Screening

D.V. Coleman

Department of Cytopathology, St. Mary's Hospital, London W2, UK

In 1989 the European Commission set itself the target of reducing mortality and morbidity from cancer in the member states of the European Community (EC). This was to be achieved through the establishment of the Europe Against Cancer programme which was concerned with the advancement of education and training in cancer diagnosis, and the general promotion of cancer prevention programmes in the member states of the EC.

One of the key areas of concern of Europe Against Cancer was the efficiency and effectiveness of the cervical cancer prevention programme in the EC. Investigation of cervical cancer screening in member states showed that there was great variation in the level of organisation of the programme. Few countries offer national or regional programmes and in most screening is opportunistic (Table 1). It was also noted that the standards of training and education of medical and paramedical personnel participating in cervical cancer screening was very variable. Only five of the 12 member states had a nationally or regionally recognised training programme for non-medical or medical personnel undertaking the analysis of cervical smears and only two countries had a nationally or regionally recognised test of proficiency or qualifying examination for cytology screeners.

In an attempt to improve the situation, the European Commission, through Europe Against Cancer, agreed to support a project which had as its aim "The formulation of basic training programmes for medical and paramedical personnel participating in cervical cancer screening in the member states of the EC". The project was launched in April 1990 and is widely known as the "European Community Training Programme in Cervical Cancer Screening (ECTP.CCS)". A working party was formed composed of two representatives from every member state (one pathologist and one cytotechnologist) and six invited experts. The representatives were nominated by their national cytology societies. The representatives for Germany were Dr. U. Schenck and Mrs. I. Brake. The secretary to the

Recent Results in Cancer Research, Vol. 133

Table 1. Cervical screening in selected European countries

Country	Organisation and coverage	Ages covered (Years)	Screening frequency (Years)
Belgium	No national programme – regional service	35–60	3
Denmark	No national programme – ten local services	20–60	3
France	No national programme – opportunistic screening	–	–
Germany	National programme – no call or recall	Over 20	1
Italy	No national programme	40–70	3
The Netherlands	National programme in place	35–54	3
Portugal	No screening programme	–	–
Spain	No screening programme – pilot in one region	Over 30	2
Ireland	No screening programme	–	–
Greece	No screening programme	–	–
United Kingdom	National programme	20–64	5

Table 2. Statements prepared by ECTP

Statements relevant to cytotechnologists participating in cervical cancer screening
1. Statement on skill levels (Appendix A)
2. Statement on training requirement (Appendix B)
3. Syllabus for training in cervical cytology (Appendix C)
4. Description of aptitude test in cervical cytology (Appendix D)
5. Constitution of ECTP.CCS examination board for aptitude test (Appendix E)
6. Recognition of centres offering training programme (Appendix F)
7. Composition of ECTP.CCS training centre accreditation board (Appendix G)

Statements relevant to anatomo-pathologists participating in cervical cancer screening
1. Statement of responsibilities (Appendix H)
2. Statement on training requirement (Appendix I)
3. Description of aptitude test in cervical cytology (Appendix J)
4. Constitution of ECTP.CCS examination board for aptitude test (Appendix K)
5. Recognition of centres offering training programme (Appendix L)
6. Composition of ECTP.CCS training centre accreditation board (Appendix G)

ECTP.CCS working party was Mr. G. Douglas and Prof. D.V. Coleman (St. Mary's Hospital Medical School, London W2) is the chairman and director.

The ECTP working party has met on six occasions and has drawn up proposals for training and proficiency testing of pathologists and cytotechnologists involved in cervical cancer screening (Table 2). The proposals have been agreed by the national cytology societies of the following member states: United Kingdom, Belgium, France, the Netherlands, Germany, Denmark, Spain, Portugal, Greece, Italy and Ireland.

The proposals[1] contain statements relevant to levels of skill which should be achieved by cytotechnologists who undertake cervical screening, together with guidelines for their training and aptitude testing (Table 2). They also contain statements on the responsibilities, training and aptitude testing of pathologists who undertake the reporting of cervical smears. In addition, they set out standards for training schools and contain proposals for their accreditation.

Members of the working party have made it clear that their views on training in cytology extend beyond the remit of the ECTP.CCS and have made the following important points in the introduction to the proposals:

1. The guidelines for training in cervical cytology represent only part of the training of cytotechnologists who should also receive instruction in general cytopathology.
2. The cytotechnologist should always work under the supervision of a cytopathologist.

It is emphasised throughout that the adoption of the proposals for training in cervical cytology and aptitude testing is entirely voluntary and is a professional decision which can be made at any level: individual, regional and national levels. However, the ECTP proposals for training and testing represent only a basic minimum requirement and are intended as a guideline where none is in place. The ECTP acknowledges that in some member states national or regional requirements may be set at higher levels and supports such arrangements. However, it is gratifying to see that several countries have already made use of the information contained in the proposals. In Portugal they have been used to set standards of training in cervical screening. A large number of training schools have applied for accreditation. Three centres (one each in Denmark, the Netherlands and Germany) have agreed to pilot aptitude testing later this year. A small working party set up under the auspices of the ECTP.CCS is preparing to run the first examinations.

One of the most rewarding exercises undertaken so far by the ECTP.CCS has been the formation of a panel of cytologists whose remit was to determine "equivalent" terminology so that the aptitude test could be conducted in any language. This exercise has now been completed successfully and in itself represents a singular advance in cervical cytology screening in the EC. By agreeing "equivalent terminology", a uniform standard of quality control and a single programme of cytology audit will be possible throughout the EC.

It is important to remember that the ECTP.CCS working party is an ad hoc body set up to formulate training programmes and set quality standards

[1] A copy of the proposals can be obtained from Mr. G. Douglas, Secretary, ECTP.CCS, Department of Cytopathology, St. Mary's Hospital, Praed Street, London, W2, UK.

in cervical screening in the EC. It is anticipated that after 1993 its work will be continued by established professional organisations and extended beyond the EC. Overtures have been made to the European Federation of Cytology Societies (EFCS) with this in mind, with the intention that the EFCS take on the responsibility for aptitude testing of cytotechnologists and accreditation of training schools. The ECTP has also been in dialogue with the anatomopathology subcommittee of the UEMS in view of their mutual interest in standards of training in cytopathology.

The ECTP.CCS expects to conclude its programme with a conference on "Quality Standards in Cervical Cancer Screening" in Madrid which is being organised at the request of the Europe Against Cancer programme. This will be held 1993 and will provide a forum for discussion of the proposals of the ECTP.CCS and an opportunity to seek consensus for them from professional, government and academic bodies throughout the EC. Widespread adoption of the ECTP.CCS proposals throughout Europe over the next few years could undoubtedly lead to improvement in the efficiency and effectiveness of cervical cancer screening in the EC. It would also represent a significant achievement for the Europe Against Cancer programme.

The Austrian Regulation

J.H. Holzner

Institut für Klinische Pathologie, Währingergürtel 18–20, A-1090 Vienna, Austria

In order to understand the cytology regulations in Austria, issued in 1985, it is necessary to know something about the development of cytology in Austria.

The first applications of cytology as a diagnostic method in Austria were published by Schauenstein in the 1920s, by several hematologists (Fleischhacker, Klima) in the 1930s, and by some pathologists – mainly on bronchial cytology – in the 1940s. A broad introduction of this method, based on the publications by Papanicolaou and Traut, started after the Second World War. In the beginning it was difficult to equip cytologic laboratories because the method was not paid for by the insurance companies. It was the privilege of the Austrian Cancer Society to provide the financial basis for the equipment of the first cytologic laboratories in our country and to pay for the smears through almost 1 decade. When I was appointed head of a gynecopathologic laboratory at the University Hospital in Vienna in 1963, an explosion in the number of cytologic smears of the genital tract started. Between 1963 and 1970 the number of smears in Austria increased from 50000 to 1.2 million.

An investigation in 2000 cone biopsies which we collected during this period revealed a relatively high percentage of negative cones. This was due to a high number of false-positive smears. As most of the false smears were reported from a limited number of laboratories, this must have been due to the improper quality of the cytology laboratories involved. On the basis of this information, we contacted our health officials and succeeded in convincing them that it was necessary to issue regulations in order to improve the quality of cytology. At the same time the Austrian Society for Applied Cytology was founded on the initiation of Professor Navratil, at that time chairman of the Gynecologic Clinic in Graz. The majority of the founding members were clinicians, mainly gynecologists and pulmologists, but only few pathologists. One of the most urgent activities of the Society was to set up a training program for cytologists and cytotechnicians.

Recent Results in Cancer Research, Vol. 133

The first problem we were confronted with was the general license requirement for medical doctors in Austria. After finishing medical school, a postgraduate training period in a hospital is obligatory in order to get the license for practice. This postgraduate training includes 3 years for general practitioners, 2 years for dental medicine, and 6 years for any speciality. For instance, in pathology the 6-year postgraduate training includes 1 year of clinical medicine, 4 years of pathology, and 1 year optionally. It is also possible to obtain an additional speciality degree (*Additiv-Facharzt*) for certain fields. For this an additional training period of 3 (in special cases 2) years is necessary for formal reasons.

Because of the relatively large number of clinicians involved in cytology – and this was a rather inhomogenous group, we decided that the only way was that of an additional degree. The officials agreed to this solution and issued the following regulation:

1. The additional speciality degree for cytodiagnostics can be obtained by pathologists, specialists for laboratory medicine, pulmologists, urologists, and gynecologists.
2. Pathologists and specialists for laboratory medicine can obtain an additional degree for the whole field of genital and extragenital cytology, gynecologists for gynecologic cytology exclusively, pulmologists for cytology of the respiratory tract exclusively, urologists for urological cytology, etc.
3. The additional training period includes 1 year in histopathology and 2 years in cytology. Only pathologists are exempted from the year of histopathology because this is included in their basic training program for pathology. The training period in histopathology has to be performed in a department of pathology, licensed for postgraduate training in pathology, the training period in cytology in a specialized laboratory, headed by a cytologist and acknowledged as a training laboratory. The training program depends on the basic speciality as mentioned above.
4. The requirements for training institutions are as follows: (a) *histopathology*: department of pathology of a university hospital or of other hospitals licensed for postgraduate training in pathology; (b) *cytology*: licensed laboratory, headed by a cytologist with an additional degree, and with a minimum of 20000 gynecologic smears and/or 3000 extragenital smears per year. Training institutions are the departments of pathology of the three universities (Vienna, Graz, Innsbruck) and their subdepartments, the pathology departments of the large city hospitals and the provincial hospitals, and some special laboratories in some cities such as Klagenfurt and Salzburg.

These regulations were issued by the Federal Department of Health in 1985, after a general agreement had been reached 1 year previously. The regulations influenced the development of cytology in Austria positively. By now 113 persons have obtained the additional degree of cytology in Austria, the majority of them – about 80% – pathologists, the rest gynecologists, laboratory people, and very few pulmologists. But a shift to pathology can be seen more and more. Almost all applications in the last few years have come from pathologists.

It may be of importance to know that the relation between pathologists and clinicians has changed completely in the last 20 years. There is a very close collaboration with regular, continuous conferences between the different specialists in pathology and their clinical counterparts, as between hematopathologists and hematologists, gynecopathologists and gynecologists, etc.

The additional degree for cytology is necessary, too, to run a private cytologic laboratory and to get a contract with the health insurance companies. But it is not necessary for a gynecologist who is doing cytology exclusively for his own gynecologic patients. In this case, the reading of the smear cannot be charged.

Unfortunately, the regulations for cytotechnicians are not so clear. In official hospitals (public hospitals, university hospitals, provincial hospitals, city hospitals) only technicians who have graduated from a school of medical technology (on average with a 2–2½-year training program) may be engaged. In this training program some basic cytology is included but not enough for a practical assignment. Therefore we introduced special courses for cytotechnology in Vienna and in Graz. These courses are not offered continuously, but depend rather on the number of applicants. They include a 3-month full-time training, starting with theoretical lectures and including a predominant part of practical training in cytology laboratories. The course is finished with a theoretical and practical examination, and the graduate receives a certificate. Unfortunately, owing to trade union regulations, the obtaining of this certificate cannot be honored by a higher salary in a public hospital. Therefore there is not much attraction in obtaining this certificate at the moment. Many cytotechnicians who have been trained here and have reached high qualifications are leaving for other countries where they can get better positions financially. But negotiations are going on and we hope to reach a better solutions to this problem.

The Swiss Regulation for the Specializations of Cytopathology and Gynecocytology

J. Jenny

Zytologisches Laboratorium, Ottenhbergstrasse 71, CH-8049 Zürich Switzerland

The Swiss regulation for professional admission as a cytopathologist or gynecocytologist corresponds largely to those valid in Austria. In 1977, the Federal Council of Switzerland instructed the Federal Office of Social Insurance to reorganize the admission of cytologic laboratories to public health insurance. The following regulations were presented, but they are still at the prelegislative stage:

1. The laboratory must work under the professional direction and immediate surveillance of a scientifically trained and approved specialist; it has to be adequately staffed and equipped with the necessary instruments.
2. The director of the laboratory must document his/her professional training in the respective specialization. Admission to special training is open to physicians, pharmacists, veterinarians, and other persons with a diploma in scientific studies.
3. The quality of analytic performance in these laboratories must be ensured by controls at regular intervals. Every registered laboratory must be subjected to unexpected checking by the control institution at least once per year. The Federal Office nominates the qualified control institutions after proposal by the professional societies involved and defines their specific tasks and activities.

This draft for legislation prompted the Swiss Society of Clinical Cytology to look for modalities to ensure and attest adequate professional training for laboratory directors. The qualification was formulated, as in Austria, as a "specialized Gynecocytologist" added to the main specialization "gynecologist FMH," or as "clinical cytologist" added to "anatomopathologist."

With the introduction of this specialization, cytodiagnostics as well as histopathology we are defined as "specific medical examinations" requiring full 'clinical interpretation." This effectively precludes the additional professional training of non-physicians, i.c., pharmacists, vctcrinarians, or pcrsons with a diploma for any other profession in the sciences.

Recent Results in Cancer Research, Vol. 133

The conditions for acquiring the additional title "gynecocytologist" by a specialist in gynecology are as follows. The applicant must be a fully trained and approved gynecologist and must be able to document 6 months of full-time work in a registered histologic laboratory of gynecopathology, and a total of 12 months of full-time work in a registered laboratory of gynecocytology.

Registration of approved institutes for additional professional training is as follows. In gynecocytology:

- The laboratory must be placed under immediate supervision of a medical director who is an approved gynecocytologist or clinical cytologist.
- The laboratory must have a minimum influx of 20000 samples per year.
- Cytologic diagnoses must be formulated in plain text.
- All slides and records must be documented and stored in the archives for a minimum period of 5 years.
- The obligatory yearly statistics must comprise the number of incoming examinations, broken down in resulting diagnoses.
- The laboratory must have an adequately documented collection for instructional purposes.

In gynecopathology:

- The laboratory must be placed under the supervision of a medical director who is an approved specialist in pathological anatomy.
- It must possess adequate and sufficient material illustrating the whole range of gynecopathology.
- It must have an adequately documented collection in gynecopathology.

Basic Training in the Subspecialization. The applicant must have evaluated and diagnosed a minimum of 5000 gynecocytologic slides under the supervision of a teacher. The specific knowledge required in this subspecialization is:

- Hormonal cytodiagnostics
- Cytologic evaluation of changes in squamous epithelium, and in the endocervical and endometrial regions
- Bacteriology, mycology, and parasitology, as well as virology of the vagina
- Differential cytology, both prospective and retrospective
- Epidemiologic aspects of early cancer diagnostics
- Training in colposcopy
- Normal anatomy and histology of the female genital organs
- Histopathology, especially tumor pathology of female genital organs

The gynecocytologist is entitled to evaluate the *exfoliative* cytology of vagina, portio, and endometrium. Cytology of the mammary gland and ovary (*fine-needle aspiration cytology*) is reserved for the clinical cytologist.

The conditions for acquiring the additional title "clinical cytopathologist" for a specialist in pathological anatomy are as follows. The applicant must have completed his/her professional training as a specialist in anatomopathology. He/she must then undergo an additional special training of 18 months' full-time work in a cytology laboratory. Here, too, the definition and approval of institutions for professional training and the basic program in the specialty is described in every detail. "Clinical cytology" comprises the entire field of cytodiagnostics.

Continuous updating and training is an essential part of quality control. On the political level, on the financial level of public health insurance companies, and on the level of the general public, the necessity of attested professional updating and upgrading of all physicians is currently being stressed. The requirements formulated towards the professional knowledge of the physician in connection with liability lawsuits show a similar trend. The FMH is presently engaged in developing regulations for professional updating and continuing education.

Status of Gynecocytology in Switzerland

A laboratory of cytology should be placed under the immediate direction and supervision of a specialist entitled to be a "gynecocytologist" or "clinical cytologist". According to a poll of the Swiss Society of Clinical Cytology in 1988 there are currently 30 laboratories which process and evaluate material from gynecologic examinations. Of these, 24 laboratories were directed by an adequately trained and approved physician:

- Fourteen under the direction of a specialist in cytology (five gynecologists, nine pathologists); 443 232 cases were examined (56.5%)
- Ten under the direction of a physician with equivalent training in cytology (three gynecologists, five pathologists, two general practitioners; 278 169 cases were examined (35.5%)
- Two under the direction of a physician without specialized qualifications in cytology (one gynecologist, one general practitioner; 38 428 cases were examined (4.9%)
- Four under the direction of a medical assistant in cytology (male or female); 24 584 cases were examined. These laboratories are partly supervised by physicians on a part-time system

Ten of the 30 laboratories were headed by qualified cytologists who had not been able to acquire the additional title "specialist in gynecocytology" or in "clinical cytopathology" as they were not previously approved as specialists in gynecology FMH or pathology FMH.

Since there are still no legal regulations for the payment of fees for cytologic examinations depending on the direction of the laboratory by a "scientifically qualified" specialist physician, it is not unexpected that there are still two laboratories headed by a nonspecialist physician, and four by a medical assistant in cytology. They had acquired permission to set up the laboratories before 1984.

Up to 12 years ago, cytologic examinations were remunerated as "laboratory work", and so fees were paid even to non-physicians. Today, the individual steps of a cytologic examination as listed in the schedule of analyses by the Federal Office are clearly divided into a technical and a medical part. Theoretically, it should be impossible for a non-physician to send a bill for "medical" activities.

Considering that in 1988, an overall 92.0% of all cases were examined in laboratories under the direction of an adequately qualified physician, the present status appears acceptable.

Any approved laboratory of cytology should deal with a minimum of 20000 cases per year in order to gain a suitable scope of experience:

- A total of 710909 cases (90.6%) were in fact examined in 18 large laboratories with a yearly input of 20000–80000 (among them one laboratory directed by a non-specialist doctor with 30335 smears).
- A total of 73505 cases (9.4%) were examined in 12 laboratories with fewer than 20000 cases per year (among them five laboratories directed by a non-specialist or by a medical assistant in cytology).

In fact, only 9.4% of all smears were examined in laboratories that are not regularly serving an input minimum of 20000 per year, according to the concept of the Swiss Society of Cytology. This can only be avoided by strict legal regulations as proposed in 1977.

Another question raised in this context calls for the distinction of state and private laboratories of cytology. Among the 30 laboratories registered in the 1988 poll, 15 are state institutions (cantonal hospitals, university clinics, institutes of pathology) and 15 private: 32% of all smears were evaluated in a state institution, and 66.3% in a private institution.

Specialization of Laboratory Directors

Previously, gynecocytology was mainly performed by gynecologists. According to the 1988 poll, the proportion of smears per supervising physicians was:

46.6%	Gynecologist
41.3%	Pathologist
9.0%	General practitioner
3.1%	Medical assistant in cytology

Although the subtitle "specialist in gynecologic cytology" will ensure that gynecologic cytodiagnostics can remain the responsibility and within the competence of the gynecologist, there is presently a marked trend towards locating gynecologic cytodiagnostics in the institutes of pathology. The actual cause may be the lack of well-motivated younger colleagues. Early diagnostics of gynecologic cancer lead a kind of shadow life in many institutions of professional training, having become a mere routine obligation. Diagnostics in gynecologic morphology have been transferre from the gynecologic clinics to the institutes of pathology. Except in two cantonal hospitals, there is no Swiss clinic of gynecology where the cytology laboratory is placed under the direct responsibility by the clinic's director.

Although this statistical compilation covers no more than some 85% of all gynecocytologic examinations, the recent activities to improve quality control have not been fruitless. All in all, the current status is far from ideal, but satisfactory.

Diagnostic Cytology in the United States

L.G. Koss

Montefiore Medical Center, Albert Einstein College of Medicine, Bronx, NY 10467, USA

The situation of diagnostic cytology in the United States is rather simple because, as cytology developed in the United States after the initial few years, it became the domaine of pathologists and not of clinicians. Therefore the standards for performance of cytology have been established by the American Board of Pathology and not by government agency. Current requirements for special certification in cytopathology require training of 4 years in anatomic pathology, with a certificate in anatomic pathology, plus 1 year of specialized approved fellowship in cytopathology that allows the candidate to sit for a special examination in cytopathology. In other words, one needs the diploma in anatomic pathology first to be approved as a specialist plus 1 more year of approved training to obtain certification in cytopathology. Other pathologists of course are not prevented from practising cytopathology but they must be certified in anatomic pathology before they can practise.

On the other hand, we have a number of regulations as part of the so-called clinical laboratory improvement act which was first passed in 1968 and had been updated repeatedly. Currently the requirements for a cytology laboratory will call for a twice yearly practical examination of all cytology laboratories by means of actual slides. Examiners may come to any laboratory performing cytologic studies with a set of unknown slides; these slides will have to be reviewed on the spot by the cytotechnologist and the pathologist responsible for it. There are some other limitations in the volume of work that can be performed by a cytotechnologist imposed by the Congress of the United States, largely because of the failures of the screening system and the failure of the society to police itself. That really is the crux of the matter, because today, if a woman dies of cervical cancer after previous alleged negative cytologic screening, the laboratory is going to be in deep legal trouble. Ultimately, if negligence is proven, as is very often the case, a great deal of moncy will changc hands. Although I do not necessarily approve the system as the optimal way of achieving excellence in any diagnostic discipline, I can assure you that the fear of a $5-million dollar fine for negligence is a very strong motivation.

Cytotechnologists in Germany

J. Lange

Lehranstalt für Zytologie-Ass., Moorenstr. 5, 40225 Düsseldorf, Fed. Rep. of Germany

Around 1965 the annual screening program for cervical cytology was established. Very soon qualified technicians had to be trained. A proposed advanced training program for medical technicians was not successful. Between 1967 and 1975 eight schools for cytotechnologists were founded, first on a 1-year training program, but soon it was extended to a 2-year training course. Besides histology and pathology, the training now included all aspects of clinical and diagnostic cytology, genetics, hematology, bacteriology, cytochemistry, and so on. However, there are no federal regulations in Germany to train cytotechnologists. State regulations only exist in North Rhine-Westphalia, Bavaria, and Lower Saxony. Cytotechnologists of the five new eastern states are medical technicians who were trained for cytology in service after a previous 3-year training period as medical technicians. After passing an examination, they are qualified medical technicians in cytology.

The *Verein Deutscher Cytologie-Assistenten* (VDCA), a professional organization founded 1973, made inquiries among its members regarding the working conditions for cytotechnologists: 57% are working in a private laboratory, 11% in university facilities, 25% in other hospitals, 3% in different institutions, and 5% are independent assistants.

One question was how many slides – not cases – a cytotechnologist sreened per day since often one case consists of more than one slide. Compaired to the United States, where sreening is limited to 80 slides per day and person, no similar regulation exists in Germany. In North Rhine-Westphalia two thirds of the technicians screened more than 100 slides per 8-h working day with a maximum of 170 slides. In Bavaria the maximun was 100 slides and in Baden-Württemberg 130 slides screened per day. These numbers will most likely increase within the next few years because of a foreseeable shortage of well-trained cytotechologists when one considers that 7 million gynecological cases have to be screened per year.

Recent Results in Cancer Research, Vol. 133

When one looks at the rescreening of random samples of negative gynecological preparations by a medical cytologists, only 43% are checked. In some laboratories even the positive slides are signed out without rescreening. This is one of the explanations given when cytotechnologists complain of being left alone with diagnostic problems and that there is no time and no one with whom to discuss the suspicious cases or positive slides. Nearly 10% have an opportunity to discuss the cases once a month, or even less in the case of a medical cytologist. That means that motivation towards a high standard is missing in this group.

A total of 20% of all independent cytotechnologists have no feedback at all. This group amounts to 5% in our opinion poll; they are employed by gynecologists paid per case. Usually these are well-trained cytotechnologists who screen the slides at home. That does not mean that they are responsible for the diagnosis nor that they are really independent. Qualified cytotechnologists should have the opportunity to screen at home, but simultaneously there has to be a guarantee that the responsible gynecologist is well trained in cytology and that the positive and suspicious slides are rescreened by him or her, classified, and then discussed with the technician. The medical professional is responsible for the diagnosis and he or she has to supervise the technician.

What about salary? For many years the VDCA has been working towards an adequate salary for cytotechnologists. The entry level for a 21-year-old technician working in public institutions is DM 2300 per month. If she is 43 years old the salary amounts to DM 3400 monthly. This salary is still based on a 1-year training program although all schools have a 2-year training program today.

What about the cytotechnologists who work in private laboratories? More than 90% earn less than DM 3500, and more than 50% less than DM 3000 depending on the different states of the Federal Republic.

After some years working as a cytotechnologist many of them change their profession or go to university again. Some of the reasons are:

- No professional advancement
- Highly qualified work with no variation
- Screening 6–8 h
- No feedback
- No adequate salary

The paucity of well-trained cytotechnologists, because of the limited number of students per year and because quite a few take up a different job, leads to the fact that cytology is being taught in so-called crash courses. This must cause a devaluation in the position of the cytotechnologist and cytology in general.

In order to revalue the cytotechnologist it is absolutely necessary to get federal regulations for a training program for cytotechnologists, especially in view of a united Europe.

The amendment to the regulations for training medical technicians (MTA) will give a unique opportunity to integrate training in cytology into the training of medical technicians. After a 1.5 years of basic medical training a separation should take place between laboratory technicians and specialized technicians in morphology including cytology. These technicians would be able to work in a cytology as well as in a histology laboratory.

The eight training centers for cytotechnologists in Germany are not able to train as many technicians as needed. Approximately 50 qualified cytotechnicians finish school every year. With integration into the program for the medical technicians, more cytotechnologists could be trained. But the professional organization of the medical technicians as well as the Minister of Health defeated this recommendation which we strongly proposed.

It is obvious that we need more training centers for cytotechnologists and at the same time a federal training and examination program. Cytological preparations should be screened only by well-trained specialized cytotechnologists and not by unqualified personnel from a different background.

To reduce the necessity of repeat smears – often a sign of insecurity – and to prevent misdiagnosis especially in gynecological cytology, we are in urgent need of highly specialized and motivated cytotechnologists.

Outlook for Diagnostic Cytology in Germany

H.-A. Müller

Pathologisches Institut der Universität Würzburg, Joseph-Schneider-Str. 2,
97080 Würzburg, Fed. Rep. of Germany

We have shown in the previous article (Lange, this volume) that the situation of diagnostic cytology in Germany is rather disquieting and somewhat shameful. The unfortunate conditions are due to several deficiencies in personnel, structure, and material equipment in educational institutions, in institutes of pathology, and in clinics of gynecology and obstetrics. A damaging factor is seen in the strict decentralization of cytodiagnostic services in the field of prophylactic screening which is liable to continual deterioration in results. Last but not least, the criminal activities of mountebank cytologists have damaged the reputation of German cytology and cytologists in the international scientific community. Any attempt to improve and reestablish the status will have to start from the basis of the problems cited above.

Referring again to Karl Lennert's (1973) presidential opening address to the 57th Meeting of the German Society of Cytology (DGZ) (held in Karlsruhe together with the corresponding Swiss and Austrian societies), I want to cite his programmatic proposals for the establishment, staffing, and equipment of institutes of cytology:

> The establishment of (separate) departments of cytopathology is an urgent necessity since we need institutions for competent education and propagation of cytological knowledge. A limiting factor lies admittedly in the low number of available cytopathology teachers with sufficient experience and scientific foundation to head and guide such departments. We shall have to make every effort to promote the professional qualification of cytopathologists in order to vanquish the present problems.

In 1987, Holle also stressed the necessity to establish cytopathology as a separate subdiscipline for research and practical purposes, especially in the important academic institutions (Holle 1987) "This alone will help to supply the expected number of well-trained cytopathologists, and to provide a reliable theoretical foundation for the growing rate of practising diagnos-

Recent Results in Cancer Research, Vol. 133

ticians". Consequently, there have been suggestions and proposals to authorize pathologists well schooled in cytology for "diagnostic cytology" as a separate discipline by means of academic habilitation according to the German system. The idea was actually carried through once.

The absolute prerequisites for the establishment of a cytopathology department in a conventional institute of pathology are as follows. There must be conscientious and regular training in cytopathology in the framework of professional further training for specialists in pathology. Another indispensable factor is the willingness of the institute's head to ostracize any attestations by "routine" or "favor".

Within the institutional framework, cytopathology should enjoy the widest possible independence in terms of organization, personnel, and finance, but close communication must be ensured in terms of the exchange of findings and data, and the continuous rotation of residents. Truly effective training in cytopathology requires a minimum of 6 months' full-time work in cytodiagnostics for the individual rotation phases of each resident.

The system proposed for the cytopathology part of further training in pathology must be appropriately adapted for training in the morphologic departments or laboratories of gynecological clinics involved in extensive further training of gynecologists. In analogy to the conditions in pathology, this absolutely calls for an expert in exfoliative cytodiagnostics to act as head of the respective department or laboratory. Appropriate guidelines and regulations for these departmental heads may be found in the Guidelines for Cytologists issued in 1980 (Richtlinien 1980). Actually this means that any larger gynecology and obstetrics hospital must have a budget-approved ward physician (morphologist) or a similarly competent senior physician to act as head of the cytodiagnostics department.

It should be, or become, unthinkable that the cytology laboratories of academic clinics run their cytodiagnostic services – even if only intramural – without a head physician of the department. Even worse, such laboratories must never employ reference gynecologists or pathologists without special training and qualifications who can only serve as incompetent and disreputable figleaves.

The second set of problems mentioned above concerns the severe decentralization of cytodiagnostic services in Germany resulting in less than optimal medical training of physicians, and eventually in a lower quality of diagnostic data. Without any doubt, the consequences of decentralization are directly contravening the original security concept of the KV. Moreover, it should not be overlooked that a larger population of less than optimally trained diagnosticians with a minor or minimal output of investigated material is practically more liable to succumb to unscrupulous activities. There is an obvious solution, but it is rather simplistic and profoundly unrealistic, and will not be mentioned here.

Nevertheless, one principle is undisputed in the attempts to solve the problem: the current situation of gynecologic cytodiagnostics may be improved by stopping and reducing decentralization, and by raising the output of diagnostic smears in the appropriate clinics and institutes, thereby reducing the number of suboptimal diagnosticians and the financial attraction of unscrupulous practice, and improving the training of future cytologists.

Presently, certain trends are favoring a compromise in the controversial opinions of the KV and partners interested not only in the application, but also in research and teaching of cytodiagnostics. Such trends have been included in the new federal comprehensive contract for physicians; but the relevant regulations were adopted and implemented only by the KV of Hesse, Rhineland, Schleswig-Holstein, but not of the other federal states. Some paradigmatic regulations have come to our knowledge, convened in Bavaria between academic institutions and practising pathologists, but so far they have not attained the force of exemplary signals.

Serious consultations are needed regarding the possibility of establish several central institutions in Germany which, based on a large volume of input material and a core of competent and experienced cytologists and cytopathologists, could be authorized to act as academic schools of training and clinical research. Such institutions might very well be situated in the "new" federal states where some remaining institutes and clinics would meet, on account of a large body of material, such requirements more aptly than in the former West Germany where the appropriate institutions have to be installed, probably at considerable expense.

This is the place to mention financial considerations in context with the above problems of applied cytology. To ignore them might be interpreted as pusillanimity or rose-colored lack of realism. There is hardly any doubt that gynecological cytodiagnostics in particular is a source of abundant additional remuneration, as reflected in the mocking term "cytology on the side" often applied to practising gynecologists with a full normal workload.

The transfer of cytodiagnostic services from the practising gynecologist (or other doctor with cytodiagnostic activities) to a larger institute or clinic will certainly meet strong opposition from the doctors involved and their professional representatives, not only on account of possible financial loss, but also because of the expected gain of the institutional counterpart. In order to obviate these certainly considerable obstacles on the way towards improved cytodiagnostics, there is an indication to develop suitable models to disarm the expected opposition. For instance, the examination of cytologic slides (smears) could become the principal task of appropriately competent clinics or institutions, instead of being performed as a mere job on the side; the potential fees could be transferred, after deduction of costs, to certain foundations or non-profit-making institutions.

Regulations of this kind would presume a lot of goodwill on the part of all the people involved, they call for solidarity in all those concerned in cytology as a scientific obligation. With the words of a nestor of European cytology, Paul Lopez Cardozo, preceding his *Atlas of Clinical Cytology* (1975): "Cytology – love it or leave it".

In view of the dramatic status of applied cytology in Germany, the German Society of Cytology feels called upon to eliminate the above inefficiency and failure as best it can. It will support the definitive implementation of the principles outlined in the recent comprehensive contract wherever this has not yet been possible for whatever reasons. It will further endeavor to establish an efficient quality control of cytodiagnostic services in Germany. At the same time the society will confront any unscrupulous activities in cytodiagnostic practice.

Above all the German Society of Cytology will engage in an indepth and extensive discussion of all the problems of applied cytology in Germany. A meeting is planned on the structural and personal development in cytology; invitations will be tendered not only to experts in cytology, but representatives of other medical societies and professional organizations, of the federal and regional chambers of physisians and KVs together with officials from the educational and public health administrations and of the insurance services. The German cytologists concerned about their specialization sincerely hope that this concerted action will help to save cytomorphology as a distinct and honorable field of science.

References

Lennert K (1973) Begrüßungsansprache. Verh Dtsch Ges Pathol 57:24–26
Lopes Cardozo P (1975) Atlas of clinical cytology. Verlag Chemie, Weinheim
Holle G (1987) Diagnostische Zytologie – Zytopathologie. Pathologe 8:132–136
Richtlinien der Kassenärztlichen Bundesvereinigung über die Voraussetzungen zur Durchführung von zytologischen Untersuchungen im Rahmen der Krebsfrüherkennungsmaßnahmen bei Frauen (Zytologie-Richtlinien) (1980) Dtsch Arztebl 77:154

Subject Index

Springer-Verlag and the Environment

We at Springer-Verlag firmly believe that an international science publisher has a special obligation to the environment, and our corporate policies consistently reflect this conviction.

We also expect our business partners – paper mills, printers, packaging manufacturers, etc. – to commit themselves to using environmentally friendly materials and production processes.

The paper in this book is made from low- or no-chlorine pulp and is acid free, in conformance with international standards for paper permanency.

Printed by Publishers' Graphics LLC